MAKING MATHEMATICS MORE PRACTICAL
Implementation in the Schools

MAKING MATHEMATICS MORE PRACTICAL

Implementation in the Schools

Editors

Leong Yew Hoong, Tay Eng Guan
Quek Khiok Seng, Toh Tin Lam
Toh Pee Choon, Jaguthsing Dindyal
Ho Foo Him and Romina An Soon Yap

Nanyang Technological University, Singapore

W🌐 World Scientific

NEW JERSEY · LONDON · SINGAPORE · BEIJING · SHANGHAI · HONG KONG · TAIPEI · CHENNAI

Published by

World Scientific Publishing Co. Pte. Ltd.
5 Toh Tuck Link, Singapore 596224
USA office: 27 Warren Street, Suite 401-402, Hackensack, NJ 07601
UK office: 57 Shelton Street, Covent Garden, London WC2H 9HE

British Library Cataloguing-in-Publication Data
A catalogue record for this book is available from the British Library.

MAKING MATHEMATICS MORE PRACTICAL
Implementation in the Schools

ISBN 978-981-4569-07-1

Printed in Singapore

Preface

In 2009, we started a research project known as Mathematics Problem Solving for Everyone (MProSE). As the title suggests, the vision was (and still is) to integrate problem solving into the regular teaching of mathematics in Singapore schools – and not just for a small select segment of the student population in the schools, but *for everyone*. We knew that the work involved would be "hard and unglamorous" – a phrase we have oft-quoted from Schoenfeld (2007). We are glad that we are still right in the thick of the work now, and with encouragements of positive experiences from teachers and students along the way.

In 2011 we produced a book entitled *Making Mathematics Practical*, which detailed the theoretical foundations of the project and a set of materials – including problems, solutions, lesson guides, and assessment tools – that teachers can use to carry out the MProSE problem solving lessons. The use of the word *Practical* has these intended meanings: (1) we speak of a Practical Paradigm of doing mathematics – focusing on mathematical processes – that is analogical to the Practical sessions in science lessons; (2) in the sense that we intend to make the MProSE design work as a mainstay in actual mathematics classrooms.

The ideas and resources in *Making Mathematics Practical* were indeed adopted and adapted by a number of MProSE participating schools. We think that the experiences of the teachers in the process of implementation and the adjustments to MProSE features as part of tweaking in a Design Experiment may be useful to others who are looking into our project from the 'outside' (and perhaps considering to come 'inside'). This is the motivation for *Making Mathematics more Practical: Implementation in the schools.*

For readers who are unfamiliar with MProSE, Chapter 1 is intended to provide an overview of the entire setup of MProSE – its theoretical underpinnings and its design framework. Chapters 2 through 6 contain reports from five schools on how they carry out the MProSE problem solving module in their respective schools. We think that teachers who are interested to know how MProSE may work in their schools and classrooms would find these chapters particularly helpful as they contain details on the benefits and challenges in implementation. That the chapters are authored by teachers from a wide range of schools in Singapore increase the likelihood that all teacher readers can resonate with some of the descriptions in these chapters. In each of these chapters, a member of the MProSE research team – one who worked most closely with the school – added their comments alongside or in response to the issues raised by the teacher authors. The final chapter brings the information about the project – from the perspective of the researchers – to its currency by including updates of the second round of implementation in the schools.

We like to record a word of thanks to Tabitha Liu for her help in editing the earlier drafts of this book. Her effort and expertise were invaluable.

Editors

Table of Contents

Chapter 1

Mathematical Problem Solving for Everyone: Infusion and Diffusion

TOH Tin Lam   TAY Eng Guan   LEONG Yew Hoong

QUEK Khiok Seng   HO Foo Him   Jaguthsing DINDYAL

TOH Pee Choon

Mathematical problem solving as the centre of mathematics learning has been the vision of the Singapore school mathematics curriculum since the 1990s. Enacting it in a school curriculum has remained a difficult enterprise. Mathematical Problem Solving for Everyone (MProSE) was a design experiment project that successfully implemented a problem solving module, based on the model described by George Pólya (1945) and enhanced with Alan Schoenfeld's (1985) observations, in an Integrated Programme school. MProSE: Infusion and Diffusion (MInD) is a new project that intends to adapt the MProSE design to a number of mainstream schools. This chapter describes the theory underlying the MProSE design, its initial success under the MProSE project, and what needs to be done for the design to be diffused to the various schools in the MInD project.

1 Problem Solving—Central to Peripheral?

In Singapore, mathematical problem solving has been established as the central theme of the primary and secondary mathematics curriculum since the 1990s. The centrality of mathematical problem solving can also be seen in the pentagon framework (see Figure 1) of the Singapore mathematics curriculum. The Ministry of Education (MOE) syllabus document explicitly states the importance of problem solving: "Mathematical problem solving is central to mathematics learning. It involves the acquisition and application of mathematics concepts and skills in a wide range of situations, including non-routine, open-ended and real-world problems" (MOE, 2006, p. 2).

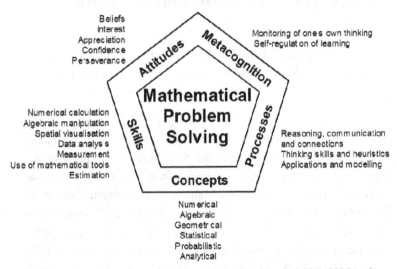

Figure 1. Singapore mathematics curriculum framework (MOE, 2006, p. 2)

Although the international comparative studies PISA and TIMSS have revealed that Singapore has achieved a high level of competence in mathematics in schools, these studies have also noted a relatively weaker performance of Singapore students in solving unfamiliar problems (Kaur, 2009). As the overarching aim of the Singapore mathematics curriculum at all levels of schooling is the development of problem solving ability, research in problem solving in school mathematics in order to support classroom practice or inform curricular policy with research-based evidence is extremely important in the Singapore context.

That problem solving is given a distinguished place in the national mathematics curriculum is not unique to Singapore. A survey of the curricula of other countries such as Australia, the United Kingdom, and the USA by Stacey (2005) revealed that problem-solving appears either as a key strand in curriculum documents or as a parallel process alongside content strands. This is indicative of the importance that the international mathematics community places on problem solving as a part of the overall development of students' mathematical competency.

Research interest in mathematical problem solving has also been kept up over the past 50 years. As reviewed by English, Lesh and Fennewald (2008), since the seminal work of Pólya's *How to Solve It* (1945), research efforts in the early years started with trialling the teaching of "Pólya-style heuristics". It then moved on in the seventies and eighties to account for the role of other complex processes such as metacognition and affect alongside heuristic activities during problem solving (Kilpatrick, 1985; Schoenfeld, 1992), and shifted more recently to weaving problem solving approaches into standard classroom instructional practice (Silver, Ghousseini, Gosen, Charalambous, & Strawhun, 2005).

While the ideals of teaching problem solving are widely shared, there remain real challenges in enacting a curriculum centring on problem solving. A number of writers question the degree of success in the teaching of mathematical problem solving. Some assert that despite decades of curriculum development, students' problem solving abilities still require significant improvement with respect to the changing demands of society (Kuehner & Mauch, 2006; Lesh & Zawojewski, 2007; Lester & Kehle, 2003). Similarly, Stacey (2005) noted that while some teachers around the world had some successes in teaching problem solving especially with more able students, she lamented that "there is still a great need for improvement, so that more students can gain a deeper appreciation of what it means to think mathematically and to use mathematics to assist in their daily and working lives" (p. 341).

Local studies in Singapore seem to support this picture of limited success in teaching problem solving. Koay and Foong (1996) found in their study that the Singapore lower secondary students did not relate problem solving strategies to real-world tasks; also, Foong, Yap, and Koay (1996) observed that a number of mathematics teachers expressed their lack of confidence in teaching problem solving in primary levels.

2 MProSE—Attempting to Restore the Ideal

As to the direction of problem solving research, Alan Schoenfeld wrote in the 2007 special issue of the journal ZDM on problem solving that the current focus should lie in translating decades of theory building about problem solving into workable practices in the classroom:

> That body of research—for details and summary, see Lester (1994) and Schoenfeld (1985, 1992)—was robust and has stood the test of time. It represented significant progress on issues of problem solving, but it also left some very important issues unresolved. ... The theory had been worked out; all that needed to be done was the (hard and unglamorous) work of following through in practical terms. (Schoenfeld, 2007, p. 539)

Our team of mathematicians and mathematics educators from the National Institute of Education, Singapore, embarked on a project, Mathematical Problem Solving for Everyone (MProSE), as an attempt at doing the "hard and unglamorous" work of realising the ideals of mathematical problem solving—as envisioned to be at the heart of the Singapore mathematics curriculum—into the daily practices of mathematics classrooms. That problem solving is currently mostly theoretical talk and not common classroom enactments is attested to by numerous local studies (see for example, Foong, Yap, & Koay, 1996; Foong, 2009).

To us, the hard and unglamorous work involved three major steps: (1) initialisation of problem solving as an essential part of the mathematics curriculum in a school at a foundational year level; (2) infusion of problem solving as an embedded regular curricular and pedagogical practice across all year levels in the school; and (3) diffusion of this innovation from one school to the full range of schools in Singapore. In each of these steps, we took a complex systems approach: instead of looming in on only one component of the system, such as curriculum, instructional practices, assessment, and teacher development, we included all these aspects in our overall design research process.

In line with design research, the initialisation stage of the process involved the creation and trialling of a design. This we did successfully with our official initial school, NUS High School. (Much work was done with Temasek Junior College even earlier but that was not formalised within the framework of a funded research project.) In that project, we redesigned parts of the curriculum and structures, including assessment practices and teacher development, within this Integrated Programme (IP) school in Singapore. [The IP schools differ from mainstream schools in that they offer a six-year programme: Year 7 to Year 12; the mainstream secondary schools offer a four-year programme which ends with the students sitting for the General Certificate Education 'Ordinary' (GCE 'O')[1] level exams.] Through a series of adjustments to the design, we developed a problem solving curriculum that is now an integral part of the school's mathematics curriculum at the foundational level.

The MProSE design comprised a problem solving module and professional development sessions to equip teachers to deliver the module. There are ten lessons in the module and they are designed in such a way that they systematically introduce students to the language and tools used in the process of mathematical problem solving. The organizing feature and tool in these lessons is the Practical Worksheet which functions both as a pedagogical and assessment tool.

Through the Practical Worksheet, students are encouraged to follow a certain model to facilitate their problem solving process. In particular, the worksheet embeds Pólya's (1945) four-stage model for problem solving, encouraging the students to understand the problem, devise a plan, carry out the plan, and check and expand. Also embedded in the worksheet are aspects of Schoenfeld's (1985) framework which identify one's cognitive resources, knowledge of heuristics, control (or metacognition), and belief systems as factors that explain one's success or failure in problem solving. A copy of the Practical Worksheet is given in Appendix A.1.

Table 1 gives an overview of the theme or topics discussed for the ten lessons designed in MProSE. Further details of the curriculum—a Scheme of Work, a series of lesson plans, pedagogical suggestions, and

---

[1] This exam is set and marked by the University of Cambridge Local Examination Syndicate (UCLES).

assessment strategy—are set out in the book we authored: *Making Mathematics Practical: An approach to problem solving* (Toh, Quek, Leong, Dindyal, & Tay, 2011). The problem solving curriculum and the research outcomes of that project are well reported in the literature. (For a full list of MProSE publications, see Appendix F.)

Table 1.

Summary of the Lessons in the MProSE module

Lesson	Lesson Theme or Topic
1	What is a problem?
2	Pólya's problem solving strategy
3	Using heuristics to understand the problem
4	Using heuristics for a plan
5	The mathematics practical
6	Check and expand
7	More on adapting, extending and generalising
8	Schoenfeld's framework
9	More on control
10	Revision
11	End of module assessment

Our new project, which we call MProSE: Infusion and Diffusion (MInD), thus builds upon this initial foundation of MProSE to scale out (Infuse) and scale up (Diffuse) the innovation to other schools in Singapore. MProSE thus began as the name of a project, now completed, but is henceforth taken to mean the design of a package to teach problem solving in the main curriculum of a school.

One of the key enablers of scaling innovations is a productive collaborative researcher-practitioner partnership. This is an enterprise that requires a careful cultivation and an ongoing re-commitment. In fact, according to Lemke and Sabelli (2008), "[t]he development of effective partnerships takes 5-10 years" (p 125). Seen from this perspective of partnership building, the MProSE project was the beginning of the process which MInD proposes to continue. In terms of this partnership building exercise, NUS High School (henceforth referred to as the initial school), recommitted to infusion under MInD. To avoid confusion of terms, we use MProSE from this point onwards to refer to both the earlier project and the current one. In addition, Temasek Junior College (another IP school) and three mainstream schools, Bedok South Secondary School, Jurong Secondary School and Tanjong Katong Girls' School, agreed to work with us as well. Their stories in implementing MProSE can be read in the following chapters of this book.

The stories at this stage are mainly on the diffusion of MProSE into the schools. The design used in the initial school needed to be adjusted as it was diffused to the other schools. On the other hand, each school had to make changes (such as reorganising their curriculum and committing to teacher development for problem solving instruction) to accommodate the problem solving design. In our conception of diffusion, this notion of scalability was closely tied to sustainability. Without sustainability, there is no foundation to build upon for scaling out and scaling up. Similarly, the goal of scaling of innovations is for sustainable improvements. Such visions of sustainable innovations require systemic adjustments. We take the view that diffusion of innovations should carefully take into consideration broad scale buy-in and commitment (by teachers, school leaders and policy makers). The stories in the succeeding chapters help us to understand what have been possible and how things can be improved, both of which are important feedback features in a design experiment.

3 Description of MProSE

A description of the conceptualisation of MProSE is given in this section. It begins with the end. Over workshops and informal discussions, we asked the question of what constituted the type of student we would like our secondary school mathematics programme to produce. The answers given would almost invariably be of the following two flavours: being able to engage independently with new mathematics (independent thinker, creative, have strong mathematics content knowledge, proficient in the use of mathematics formulas, able to apply mathematics learnt and handle 'twists' in problems); and having a positive attitude towards mathematics (enjoy learning mathematics rather than hate it, see the beauty in mathematics, enjoy doing mathematics problems). Working backwards, we conjectured that really understanding and using the processes in problem solving would achieve the first disposition, and success in solving challenging problems and its consequent transfer to success in normal mathematics content would contribute to the latter disposition. Thus, we were drawn back to the centre of the stated mathematics curriculum, i.e. problem solving. The difficulty however lay in making the frameworks of Pólya and Schoenfeld work in the classroom.

From the MProSE team members' personal classroom experience of teaching problem solving, students were generally resistant to following any model of problem solving – even the higher achieving students. The metacognitive part of the problem solving process still left much to be desired. In an attempt to 'make' the students follow through a problem solving model, especially when they were clearly struggling with the problem, a 'Practical Worksheet' was used in conjunction with mathematics "practical" lessons. Doing this, MProSE achieved a paradigm shift in the way students looked at these 'difficult, unrelated' problems which had to be done in this 'special' class. It was thus concluded that it is certainly conceivable that specialised lessons and materials for mathematics (similar to the science practical) may be necessary to teach the mathematical processes, including processes that are linked to problem solving. A problem solving model was needed, and George Pólya's (1945) model (partly because Pólya's model is already proposed in the Singapore mathematics curriculum), with Alan Schoenfeld's (1985) enhancements, was adopted as the theoretical framework.

The mathematics practical lessons introduced by MProSE brought to the forefront of consciousness the essential stages one has to go through in mathematical problem solving. Data from the initial school clearly showed that the students were able to successfully solve the problems and even expanded on them using the Practical Worksheet. The "practical" paradigm played a critical role at the initial stages in easing the students into learning problem solving, where the learner acquires a model of problem solving and develops his own habits of problem solving.

To realise this paradigmatic change in the teaching of mathematical problem solving we had to address the issue of student assessment. What we saw as the root cause of the lack of success in previous attempts to implement problem solving in classrooms was that mathematical problem solving within the school curriculum was not assessed, especially in a way that mattered to the students. Because it was not assessed, students and teachers did not place much emphasis on the processes of problem solving; students were more interested to learn the other assessed components of the curriculum. Thus, MProSE designed a formal assessment system complementing the innovative curriculum that assesses not just the product but also the processes of problem solving. It includes an assessment rubric for assigning marks to students' Practical Worksheet submissions. For the readers' reference, a copy of this rubric can be found in Appendix C. If the curriculum were a dog and assessment its tail, the painful truth is that "the tail wags the dog". Most students would study what counts in an examination. Leong (2012) interviewed five students after they had attended an MProSE problem solving course. He reported that "[t]he responses from the five students seem to indicate that making the results of the module count as part of school assessment would very likely make the students put in greater effort for the course" (p. 133).

Our work with the initial school had successfully taken them along the problem solving route to a point beyond what had been reported in the literature on problem solving in Singapore. The school has continued to use the problem solving design as a regular module for all its Year 8 students (Quek, Dindyal, Toh, Leong, & Tay, 2011). Furthermore, teachers have also begun to use the language of discourse in mathematical problem solving and have adapted their instructional methods to be more in line with the use of the Practical Worksheet (Leong, Dindyal, Toh, Quek, Tay, & Lou, 2011).

One of the hurdles to overcome involved the different conceptions of the role and purpose of problem solving in relation to the overall mathematics curriculum. In this regard, the distinctions made by Schroeder and Lester (1989) are still widely used in the literature on problem solving (e.g. Ho & Hedberg, 2005; Stacey, 2005).

- Teaching *for* problem solving
- Teaching *about* problem solving
- Teaching *through* problem solving

Instead of being bogged down by choosing one particular concept over the others, we saw each of these perspectives as offering different affordances which when left out results in missing out vital elements in the overall strategy to carry out a successful problem solving agenda. The knowledge of mathematical content—especially built up in the earlier years of schooling and continually expanded—served as a resource for problem solving; there was a need to model and explicitly teach students the language and strategies used in problem solving (about problem solving); and the problem solving approach to instruction—especially after undergoing familiarity with the processes involved—could become a means (through which) to teach standard mathematical content. In other words, each of these concepts was feasible within the project. Table 2 below shows the foregrounding of each concept in relation to the stages of the research enterprise and roles.

Table 2

Role of problem solving in relation to the MProSE project

Role of problem solving	Implementation	Initial school	Mainstream schools (Diffusion)
Teaching *for* problem solving	Ongoing in main curriculum.		
Teaching *about* problem solving	Students' early familiarisation with problem solving	MProSE module (Year 8)	Adapting the problem solving design
Teaching *through* problem solving	Students later apply problem solving approach to learning content	*Infusion* at higher level (Year 9 & 10)	

One well-known factor in the success of implementing problem solving in the classroom is the teacher (Ho & Hedberg, 2005). Teachers' professional development and beliefs are thus critical in any effort to bring about significant success in teaching problem solving and has indeed been the subject of recent research. The project undertaken by Silver et al. (2005) focussed primarily on equipping teachers to use an innovative mathematics curriculum based heavily on problem solving. The teacher participants took part in the actual solving of mathematics problems, performed case analysis of other teachers' attempts at teaching those problems, as well as completed several cycles of the Lesson Study process—"selecting a target lesson, using a structured set of questions to assist in collaborative lesson planning, teaching a lesson, and discussing their lessons with colleagues" (Silver et al., 2005; p. 290). In their project, Leikin and Kawass (2005) presented teachers with a problem to solve and followed it up with a video on how a pair of students solved the same problem—from incorrect approaches to finally solving it correctly. In both of these projects, the authors reported significant changes in the teachers' practices, such as in lesson planning and in their expectations of students' abilities, with respect to problem solving.

In MProSE, teacher development is a vital piece of the overall design (Leong et al., 2011). Prior to teachers' implementation of problem solving in the classroom, we planned to have a substantial component (lasting about six months to one year) on teacher professional development, which drew on the innovations of the earlier-cited projects, such as providing teachers opportunities to solve problems, discuss plans, and participate in Lesson Study cycles.

Any attempt to implement any novel curriculum, including one centring on problem solving, must take into account the complexities of classroom practice (Ball, 2000; Lampert, 2001). Teachers need to balance different and sometimes competing goals of teaching (Wood, Cobb, & Yackel, 1995). Teaching problem solving is thus not seen as an isolated or only goal in this project but is analysed in the realistic context of other interacting and worthy instructional goals (such as keeping to time in teaching) as teachers carry out practice in the classroom. As such, the focus in this project was not only the careful framing of the roles of problem solving in the school mathematics curriculum and the ongoing development of teachers to prepare them for the enterprise but also the workability of a heavy emphasis on problem solving in actual classroom practice.

4 Research Design

MProSE proposes to use "design experiment" (Brown, 1992; Collins, 1999; Doerr & Lesh, 2003; Middleton, Gorard, Taylor & Bannan-Ritland, 2006; Quek et al, 2011; Wood & Berry, 2003) as the methodological backbone. Design experiment arose from the attempts of the education research community to address the demands of research in real-life school settings in all its complexity. The methodology of design experiment argues for the application of multiple techniques to study a complex phenomenon in education. As such it permits the use of a slew of methods such as participant observation, interviews, videotaping, and paper-and-pencil testing to provide corroborative evidence for findings.

To Schoenfeld (2009), design experiment is built on a "design-theory dualism". In our case, it began with a theory regarding mathematics education. Based on this theory, a design is conceptualised to improve learning. The initial design is then implemented in a suitable 'real-world' context, such as the school. Research methods are then used to examine both the "accuracy of the underlying ... theory *and* the power of the intervention" (Schoenfeld, 2009, p. 3). Based on the findings of the research, refinement is made to the design which is further implemented and the implementation-research-refinement cycle is iterated until the design experimenter is satisfied. "The coherence of theory to methodology ... is of fundamental importance in the evaluation of design experiments, and is critical to explicate for any future scholar or practitioner who attempts to replicate or implement the findings of a design study" (Middleton et al., 2006, p. 18).

Schoenfeld's (2009) model of design experiment involved only the interaction between the designer and the researcher in a design-theory dualism. In our work in MProSE, we were convinced that the teachers and school that implement the design have a key role in the final design itself. Black (2009) advocates this model of design experiment which includes the designer, the researcher and experienced teachers. We proposed that, in effect, a design-theory-practice troika should always be considered for a designed package to be acceptable to the final users—in this case, the teachers and the schools. In addition, we noticed that there were distinct differences in the types of changes we made during the design process. One type of change related to what is already widely discussed in the literature on design for the purposes of developing the theory and the product—refining. However, another type of change was made to meet the realistic constraints faced in practice—accommodation. It is important in design research to distinguish the two. Figure 2 shows the design-theory-practice troika underlying our proposed design experiment for MProSE including the two types of changes we made.

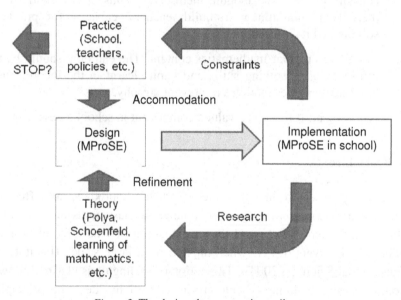

Figure 2. The design-theory-practice troika

The parameters of the MProSE design are underpinned by the theoretical justification (see Quek et al., 2011) for the design experiment. The parameters are:

1. Place in the curriculum: The MProSE design must be part of the mainstream mathematics curriculum.

2. Model of mathematical problem solving: While there are various models that can be followed for mathematical problem solving, Pólya's model (all four stages) and Schoenfeld's framework (teaching heuristics and emphasising control) are the ones adopted in the MProSE design.

3. Teachers' expertise: Teachers in school will ultimately possess the capacity to teach the module themselves. Thus, it is essential that there is a commitment to build teachers' capacity for problem solving and its instruction.

4. Relation to regular mathematics content: The end goal should be to infuse problem solving skills and habits learnt in the module into other mathematics modules to prevent atrophy.

5. Assessment: It must be a valued component in school assessment.

5 Procedure

Scaling up research involves the effort to "reproduce an effective practice in a considerably greater number of classrooms and schools" (Fuchs and Fuchs, 1998) and the investigation of "the effectiveness of education interventions … when applied on a large scale" (Institute of Educational Sciences, 2004). The notion of scaling up is clarified in two comprehensive volumes—Scale-up in education: Ideas in principle, Volume 1, and Scale-up in education: Issues in practice, Volume 2— edited by Schneider and McDonald (2007). Consisting of contributions from a multidisciplinary and interdisciplinary team of renowned researchers, these two volumes discuss the conceptualisation of 'scale-up', clarify methodological concerns, highlight issues, and present principles for successful scale-up. In the editors' words "Scale-up is the enactment of interventions whose efficacy has already been established in new contexts with the goal of producing similarly positive impacts in larger, frequently more diverse populations" (p. 5). In our case, this would refer to enacting MProSE into Singapore mainstream schools. Efforts and interests in scaling up are not new. Within mathematics education, a well-known scaling up research was done by Jim Kaput in his effort to scale up his innovation in teaching mathematics of change and variation (Kaput, 1994, 1997). Kaput was keen to find out: What can we learn from research at the next level of scale (beyond a few classrooms at a time) that we cannot learn from other sources? Scaling up research helps researchers to focus on the robustness of an innovation when used by varied students, teachers, classrooms, schools and regions (Roschelle, Tatar, Schechtman & Knudsen, 2008). Even with a single scaling up study, researchers can characterise how an innovation holds up as it spreads beyond the original trial classroom (Baker, 2007).

The diffusion of innovation was first explained by Rogers (1983) as one which is dictated by uncertainty reduction behaviour among potential adopters during the introduction of innovations. Innovative practices offer its potential adopters new ways of handling certain problems. But the uncertainty as to whether it will be better than the existing method poses an obstacle to the adoption process. To overcome this, potential adopters must seek additional information, particularly from their peers. Rogers (2003) also proposes five factors that influence the rate of adoption: observability, trialability, compatibility, complexity and relative advantage. The teachers from the initial school felt that the students could benefit from the problem solving module (observability), and that students from the entire spectrum of mathematical abilities would be able to benefit from it (relative advantage). The problems used in the MProSE were chosen to fit the school curriculum, and hence were suitable for the school students (compatibility). Furthermore, the research project aligned well to re-invention – it being a design experiment – and very clearly trialability. We also acknowledge the complexity of MProSE as a hurdle to schools' adoption. As such, we built in a programmatic approach to curricular changes and teacher development in the design, as elaborated in the next section.

According to Gorard (with Taylor, 2004), "The emphasis [in design experiments], therefore, is on a general solution that can be 'transported' to any working environment where others might determine the final product *within their particular context* (emphasis added)" (p. 101). The envisaged outcome of MProSE was to produce a workable problem solving design that could be adapted to the setting of mainstream Singapore schools. MProSE's approach in this design process was via the two main steps of diffusion and infusion.

Diffusion. This involved the diffusion of innovation to other mainstream schools. We discussed with various mainstream schools about the feasibility and constraints in implementing a problem solving curriculum. Taking into account the feedback by the principals and school teachers, we began the work of adapting from the initial design to fit the curricular and contextual demands for the participating mainstream schools.

We recognised that teachers are the key to any successful curricular or instructional reform. In the area of non-routine problem solving, past research (see, for example, Schoenfeld, 1992; Lester, 1994) have highlighted the need for teachers to experience problem solving themselves. MProSE does more than just equipping teachers with the problem solving strategies and heuristics. It introduces Pólya's (1945) problem solving model and Schoenfeld's (1985) problem solving framework to the teachers. Most importantly, MPRoSE uniquely takes the teachers through a paradigm shift by engaging them in the use of the Practical Worksheet which was crafted to focus attention on fundamental aspects of problem solving in mathematics. Teacher capacity building was done for all the new schools by early 2012.

The students from the participating schools were from the lower secondary levels. The choice of starting with lower secondary students was deliberate for the following reasons: (a) The schools were concerned that piloting the MProSE curriculum to the upper secondary students' would affect the students' preparation for the national examinations; (b) The schools had greater autonomy over the lower secondary curriculum to make changes to its enactment; (c) Building the right habits of mind with respect to mathematical problem solving has to begin as early as possible; and (d) Starting early would allow the prospect of following through to the infusion of the design at the upper secondary levels for a comprehensive curriculum change.

Infusion. From the literature on innovation diffusion, infusion is linked to local diffusion. However, we used infusion to mean the injection of problem solving techniques to learn new mathematics concepts. Our plan for infusion follows this process: (1) The students are initialised by learning about problem solving through the MProSE problem solving package; (2) The researchers modify the existing curriculum material of the higher secondary levels to engage the students to learn mathematics topics at these levels through problem solving. Stage (1) is completed for all the participating schools and details of their implementation are found in Chapters 2 – 6. Stage (2) is the next step in the MProSE research. From the research perspective, we hope to study the students' ability to apply the problem solving approach to learn new mathematics content knowledge, and how the instructional methods used by teachers could be altered to facilitate students' acquisition of mathematical knowledge.

6 Research Cycle

This section outlines the iterative cycle of the problem solving project. It consists of the following stages:

Stage 1:   Pre-research preparatory work
　　　　　(June – December 2011)

- Diffusion: Prior to the commencement of the research project, the MProSE team modified the problem solving course material that was used for NUS High School, the initial school. The team recognised that while the material used for MProSE (Toh et al., 2011) was suitable for the initial school (which is an independent school specialising in mathematics and science), the material needed to be suitably adjusted to the mainstream schools in order to diffuse the innovation to a wider spectrum of schools. This modified set of materials is henceforth referred to as the MProSE guidebook for mainstream schools. The lessons were essentially unchanged, but some of the problems were modified. The problems used in this guidebook are given in the Appendix C.

- Infusion: For infusion of problem solving, the team worked with the mathematics syllabus of the initial school to develop the material in preparation for the infusion of problem solving into their mathematics lessons.

Stage 2: Preparation for problem solving research in schools
(January – March 2012, 2013 & 2014)

This involves a series of meetings with teachers in preparing for the participation in this research project, some of which have already been carried out.

- Diffusion: The team expects to work with the teachers of the mainstream schools when adapting the problem solving module to the individual mainstream schools. Areas that may require adjustments include the actual problems used for the problem solving course, and the structures of the course subject to the constraints of the schools. Any change, however, should be made keeping in mind the basic parameters of the research: (1) The problem solving module must constitute a significant part of the students' assessment; and (2) It must be compulsory for all students within the selected level.

- Infusion: The team works with the MProSE schools to select suitable topics in the mathematics curriculum and develop new materials and lesson plans which hinge on problem solving for learning mathematics. Meetings will be arranged with the teachers for discussion on the pedagogical considerations and the individual school constraints.

MProSE researchers readily acknowledge that although embarking on the project will result in policy implications for school leaders, particularly in the area of curriculum design, it is still finally the school's prerogative to decide whether to change or not. By its insistence on remaining true to the design features while recognising a participating school's interests, MProSE's research designs must necessarily remain flexible. Its approach is interventionist. Each study within a school plays two intertwined roles: firstly, to test its hypothesised process of teaching and learning mathematical problem solving via the Mathematics Practical, and secondly, to generate new hypotheses as the study unfolds. MProSE aims to work iteratively to test and refine local theoretical models of how to carry out problem solving lessons with the particularistic contexts of each school.

An underlying purpose of MProSE's design experiment is, to borrow the words of Cobb, Confrey, diSessa, Lehrer, and Schauble (2003),

> ... to develop theories, not merely to empirically tune 'what works'. These theories are relatively humble in that they target domain-specific learning processes. For example, a number of research groups working in a domain such as geometry or statistics might collectively develop a design theory that is concerned with students' learning of key disciplinary ideas in that domain. A theory of this type would specify successive patterns in students' reasoning together with the substantiated means by which the emergence of those successive patterns can be supported. (p. 9)

In MProSE, the domain being worked on would be mathematical problem solving and the substantiated means to support successive patterns in students' reasoning would be the Pólya-Schoenfeld blend of mathematical problem solving.

MProSE's research design is thus a series of approaches in which data collection methods are decided with inputs from the schools, for example in its use of Lesson Study, to achieve a deep understanding of the complexity of the 'learning ecology' of a problem solving classroom. Its research approach engages the practitioner in the examination of the practices *in situ*, and in a co-construction of knowledge that is informed by the theoretical and practical that is situated in authentic practice. Such knowledge would be relevant research of worth to that particular school. It would enhance research utilisation and impact from MProSE's perspective, or rather from the perspective of the research enterprise.

Yet another school, with its unique learning ecology (e.g., school culture or socio-mathematical norms in the mathematics classroom), would demand an accommodation of the MProSE problem solving design. Through the gradual process which MProSE has metaphorically labelled as diffusion, the problem solving curriculum design can be accommodated to meet the salient features of each school which are keen to adopt the innovation.

MProSE's intention of working with schools which are keen on meeting the challenges of implementing the problem solving package is based on our belief that schools have their policy and practice needs that must be met or negotiated for successful research intervention and utilisation. Doing so allows the MProSE team to be sensitive to the nuanced interpretations of research problems by the school, refine its research questions, respond to contextual changes in practice, and remain valid in drawing implications for school policy and practice. In our discussions with schools following from the presentation of the problem solving curriculum package to about 250 participants, we learned that for successful implementation of MProSE's curricular innovation, it must be relevant and be of worth to the school, and that the school must show leadership support for the teachers to explore and change practices. After all, practice is shaped by a school's values and socio-cultural history, and would therefore require willingness and openness on the part of the teachers and the school leaders to allow classroom practices to be examined (for example, through lesson studies in MProSE's research design). Finally, the school's adaptation of the MProSE's problem solving design is contingent upon what the teachers themselves have conceptualised as mathematical problem solving, for example, as the application of heuristics or as a way of working and thinking for a mathematician.

For a school which has successfully integrated MProSE's problem solving package into its mathematics curriculum at a particular secondary level, the process of gradual spreading of the curricular innovation to the other secondary levels requires a constant exchange of knowledge between the researchers and the school (teachers and school leaders) to determine ways of infusing the essence of the disciplinary practice of problem solving in mathematics. MProSE hypothesises that infusion is possible through the co-construction of knowledge between researchers and teachers that in turn helps acculturate teachers, and thenceforth their students, into problem solving that quintessentially is a mathematical disciplinarity. According to Hogan (2007), disciplinarity in mathematics implies "immersion in authentic ... practices [such as] generating, validating (justifying), communicating and debating knowledge claims; deep (conceptual) understanding of key concepts and principles and their relationships (knowledge); organizing the classroom as an epistemic community [that encourages] collective reflection and discussion; and cultivating appropriate epistemic dispositions" (slide 45). An understanding of the processes within a school that has facilitated infusion is therefore essential to MProSE's endeavour to embed problem solving within the mathematics curriculum. It will form the basis for the schools which are beginning to initialise problem solving.

The following chapters will enlighten us on the perspective of the schools while they actually implemented the MProSE vision: Mathematical problem solving for everyone.

Chapter 2

Mathematical Problem Solving in Temasek Junior College

YEO Chiu Jin   HSI Han Yin   Jonathan LAU Wee Lip
HO Foo Him (NIE Researcher)

Mathematical problem solving has been part of the mathematics curriculum of Temasek Junior College Integrated Programme since 2005. This chapter describes how this aspect of our mathematics curriculum has evolved over the years. We also share various non-routine problems that we used to support students' problem solving processes. Finally, we point out some of the challenges we faced in carrying out mathematical problem solving.

1 Introduction

The Integrated Programme (IP) for academically-inclined secondary students was implemented in the Singapore education system in 2003. Under this programme, secondary students need not take the GCE 'O' Level examinations at the end of their fourth year but instead can directly proceed with their two-year Junior College education. As such, more curriculum time can be dedicated to their holistic development. This programme was introduced to create a more diverse, flexible and ability-based education pathway for different students. Initially implemented only in a handful of schools, the programme has since expanded. Temasek Junior College's (TJC) own engagement with the IP programme began in 2005. It is against this backdrop of transition in TJC that the events reported in this paper occurred.

Secondary schools conferred with an IP status normally offer six years of secondary education (i.e., Year 7 to Year 12) to the students, or it may do so in partnership with a Junior College. In the case of TJC, however, it originally provided education at Years 11 and 12 and under IP first extended the programme to Years 9 and 10 in 2005 (before finally extending further in 2013 to include Years 7 and 8).

In line with global trends in mathematics education, mathematical problem solving has been established as the central theme of the Singapore primary and secondary mathematics curriculum since the 1990s. Indeed, mathematical problem solving is at the heart of MOE's mathematics curriculum framework (MOE, 2006). Therefore, we hoped to include problem solving in TJC's mathematics IP curriculum to develop students' mathematical thinking.

Pólya (1945) described the mathematics problem solving process as one involving four main stages, namely, understanding the problem, devising a plan, carrying out the plan, and looking back. According to Schoenfeld (1985), various categories of knowledge and skills are needed to be successful in mathematics problem solving: resources, heuristics, and control. These categories of knowledge and skills are needed for students to succeed in the problem solving process. For instance, certain problems will require that students possess a certain set of resources or knowledge of specific mathematics content before the problem can be understood or solved. Heuristics, such as pattern finding, systematic listing, and creating representations, are often necessary to unlock certain problems, or form and prove conjectures. Finally, students need to possess a certain level of control or self-monitoring in order to assess whether they are pursuing productive routes in their problem solving process or not. Besides these knowledge and skills, Schoenfeld also recognised that students' beliefs in their ability to solve problems are important. Thus, students should be imparted with the correct beliefs to support them towards achieving successful problem solving.

Schoenfeld (1994) suggested that a more systematic approach to solving non-routine problems and the use of engaging practices that stretch and add value to students' intellectual development in mathematics will help in the development of students' thinking skills.

Taking these things in consideration, our school introduced mathematical problem solving that was rooted in Pólya and Schoenfeld to our IP mathematics curriculum in 2005.

2 Mathematical Problem Solving in TJC During the Pre-MProSE Years

Prior to the official commencement of the Mathematical Problem Solving for Everyone (MProSE) project, the school was already involved in mathematics problem solving. Some of the practices employed in this pre-MProSE phase, such as the use of the Practical Worksheet, form the seminal features of MProSE. After a hiatus of a few years, TJC currently rejoined the MProSE project. For the purpose of this chapter, we focus on the Pre-MProSE experience.

In those years, the college IP mathematics curriculum committee was tasked to design and develop a holistic and practical mathematics curriculum for IP Years 9 and 10. Apart from other curricular innovations, mathematical problem solving was one of the key components in the entire IP mathematics curriculum. Working closely with a research team led by Tay Eng Guan at the National Institute of Education (NIE), Ho Foo Him[1] and the mathematics department key personnel spearheaded the planning and implementation of a mathematics problem solving module that was subsequently extended to all the IP students in 2005.

Prior to the implementation of this problem solving module, a half-day workshop on Pólya's (1945) model and Schoenfeld's (1985) problem solving framework was conducted for all the teachers in the mathematics department. With strong support from the college leaders, the once-a-week problem solving module commenced in the second week of Term 2[2] during normal curricular time in 2005.

---

[1] Ho Foo Him was previously Senior Teacher at TJC and is now an NIE researcher in the MProSE team.

[2] In TJC, the school year is divided into four equal Terms, each with a duration of eight weeks.

A member of the NIE research team taught the six tutorial sessions in Term 2 of one particular class, where each session was one-and-a-half hours long. These tutorial sessions served as "model lessons" and were observed by the collaborating teachers and at least one other researcher from NIE. In each model lesson, the teacher-researcher went through some of the problems which were crafted by the research team, highlighting and illustrating how different problem solving heuristics could be used to understand and solve these problems. Figure 1 gives an example of some of the problems used in the first tutorial. After this, students worked in pairs to attempt the rest of the problems in class within a specified time.

After each model lesson, the collaborating teachers would then conduct the tutorial in their own classes on different days of the same week. A detailed account of this work was reported in Tay, Quek, Toh, Dong, and Ho (2007). In Term 3, another six sessions were conducted. This time, a resident teacher took over the tutorial sessions that served as model lessons.

---

1. There are n people in a party. Every two persons will shake their hands. How many handshakes are there altogether?

2. How many squares are there in an n x n square? What about the number of squares in an n x m rectangle, where $n > m$?

3. How many rectangles are there in an n x n square? How many of these rectangles contain an even number of unit squares?

4. How many cubes are there in an n x n x n cube?

5. How many cuboids are there in an n x n x n cube?

Figure 1. Five of the problems in Tutorial 1 of the module

The research team used different types of problems in their model lessons. Figure 2 shows a good range of problems set by the researchers. To encourage professional development, collaborating teachers were strongly encouraged to contribute problems for the module in the subsequent lessons.

---

1. Prove or disprove each of the following statements.
   (i) Every prime number is odd.
   (ii) The sum of two rational numbers is rational.
   (iii) The sum of two irrational numbers is irrational.
   (iv) The sum of two irrational numbers is rational.
   (v) The sum of a rational number and an irrational number is irrational.

2. (a) If x, y and z are positive real numbers such that $2^x = 5^y = 10^z$, show that $\dfrac{1}{x} + \dfrac{1}{y} = \dfrac{1}{z}$.

   (b) Prove a similar general result such that if $3^x = 4^y = 6^z$

   (i) show that $\dfrac{1}{2y} = \dfrac{1}{z} - \dfrac{1}{x}$.

   (ii) Estimate an integer value for p if $2x = py$.

3. The *greatest integer* in a real number x, denoted by $[x]$, is the largest integer less than or equal to x. For example, $[2.3] = 2$, $[3] = 3$, $[0.9] = 0$, $[-1.2] = -2$. $[-0.8] = -1$, etc.

   Without using a calculator, evaluate $\left[\left(2 + \sqrt{3}\right)^4 \right]$.

4. Show that if n is an integer, then $[x + n] = [x] + n$ whenever x is a real number.

Figure 2. Some varieties of problems set by the researchers

Classroom observations obtained from the twelve lessons conducted in Terms 2 and 3 revealed that most of the students generally did not know how to organise and pen their thought processes using Pólya's model. They did not employ appropriate heuristics productively. They lacked the ability to exercise or control their metacognitive skills to "free" themselves in dead-end situations in problem solving. And they did not generally make the extra effort to check and expand the problem. The latter was true even for the higher achieving students who could solve the given problems.

Although there was some success especially amongst the better students, these shortcomings needed to be examined and rectified. The research team proposed that more scaffoldings and "triggers" were required for students to apply Pólya's model and Schoenfeld's framework during the problem solving process. In addition, the research team argued that if science practical lessons were part and parcel in the teaching and learning in the sciences, mathematics should have a similar "practical" component that is similar to doing science experiments that emphasise process. As a result of discussions between the research team and the teachers, a problem solving template called the "Practical Worksheet" was specially designed to address the above limitations. Figure 3 shows a condensed form of the 2006 version of the worksheet.[3]

When the problem solving module was implemented again for the new batch of Year 9 and Year 10 students in 2006, the Practical Worksheet became a key feature of the problem solving module in that students had to use it in every lesson. In order to give students more time to engage in tackling the intricacies of the problems in class, the number of questions per tutorial session was reduced to two or three. In addition, only one of the questions was required to be answered on the Practical Worksheet. The students then submitted the filled-up worksheet after 15 to 20 minutes depending on the length and difficulty of the problem.

The problem solving module for the Year 10 students was stretched over three terms consisting of ten lessons. On the other hand, similar to the previous year, the problem solving module started in Term 2 for the new Year 9 students. For each tutorial class, a particular mathematics teacher was assigned to teach the entire module. Teachers were encouraged to contribute problems for the Year 10 classes and these

---

[3] In this earlier version of the Practical Worksheet, Pólya's fourth stage was written as "Check and Extend". In the later version, this stage was rephrased as "Check and Expand". We think that "expand" covers a wider scope such as seeking alternative solutions, adaptations, including extensions.

problems were submitted to the research team for comments before they were adopted. Figure 4 shows two problems set by one of the teachers.

With regard to each problem given to you, solve it using Pólya's Problem Solving Model and at the same time, write your method and observations on the following phases of Pólya's Problem Solving Model.

Question _____

I   **Understand the problem**
         (You may have to return to this section a few times. Number each attempt to understand the problem accordingly as Attempt 1, Attempt 2, etc.)
         (a) What is your feeling about the problem? Does it bore you? scare you? challenge you?
         (b) Is there any part you don't understand?
         (c) Which heuristics did you use to understand the problem?

II   **Devise a plan**
         (You may have to return to this section a few times. Number each new plan accordingly as 1, 2, etc.)
         Do you think you have the required resources?
         (a) What are the key concepts that might be involved in solving the question?
         (b) Which heuristics did you use to explore the problem?
         (c) Which heuristics did you use to devise a plan?
         (d) Write out each plan concisely and clearly.

III   **Carry out the plan**
         (You may have to return to this section a few times. Number each implementation accordingly as 1, 2, etc.)
         (a) Do you think you have the required resources? If you need resources ask teacher.
         (b) Do you think you are in control of the problem?
         (c) Write out each implementation in detail.

IV   **Check and Extend**
         (a) How did you check your solution?
         (b) Do you think there is a better solution than yours?
         (c) Can your method be used to solve 'similar' problems?
         (d) Give at least one adaptation or extension or generalisation of the problem.

Figure 3. Practical Worksheet (2006 version)

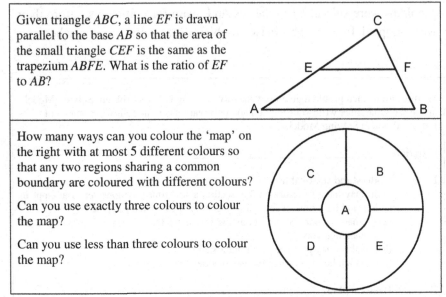

Given triangle *ABC*, a line *EF* is drawn parallel to the base *AB* so that the area of the small triangle *CEF* is the same as the trapezium *ABFE*. What is the ratio of *EF* to *AB*?

How many ways can you colour the 'map' on the right with at most 5 different colours so that any two regions sharing a common boundary are coloured with different colours?

Can you use exactly three colours to colour the map?

Can you use less than three colours to colour the map?

Figure 4. A sample of problems set by teachers

Aligned with the school's emphasis of incorporating Information and Communication Technology (ICT) into teaching and learning, all the Year 9 students had a personal tablet computer. Learning resources and notes as well as a blank Practical Worksheet were available on the learning portal, reducing the burden of printing a hardcopy for students. Thus, it was common to see students submit their typewritten work on the Practical Worksheet through the learning portal. However, the research team and the teachers noticed that many students either did not have the habit or see the importance of writing down their feelings or thought processes on the Practical Worksheet even though they were unable to solve the problem.

In 2007, seeing the importance of eliciting students' responses to the guide questions in the Practical Worksheet, Foo Him added checkboxes[4] to help students indicate their feelings and reflections about the questions attempted.

---

[4] See the sample work shown in Figure 13.

Though the problem solving module was part of the mathematics curriculum for both Year 9 and Year 10 students, the teachers were not restricted to conduct these lessons in a fixed schedule. Instead, they had the flexibility of carrying out a total of six to eight sessions anytime within Term 2 and Term 3. The level of commitment of implementing the problem solving module and the consistency of adopting the desired teaching approaches to achieve the intended objectives of the module, however, did not seem to be as strong as when it was first implemented in 2006. This was attributed to the fact that a few new teachers[5] were teaching the module. In addition, there were other education priorities for students that teachers needed to attend to such as placing greater emphasis on using ICT and service learning. Nevertheless, there were a few teachers who still upheld the practice of using the Practical Worksheet in problem solving in their own tutorial classes.

Assessment remained as one of the important driving forces that motivated students to engage in learning and achieving good results. Since a substantial amount of curriculum time was invested in teaching the module, it was only fair that students are assessed and the effectiveness of the module be evaluated. During the Pre-MProSE years from 2005 to 2007, at least 10% of the marks were allocated for non-routine problems to assess students' ability in problem solving and these problems were intentionally embedded in their final examination papers.

3 Mathematical Problem Solving in TJC during the MProSE Years

The problem solving module that was conceptualised, designed and implemented for our IP students by the researchers and the teachers at TJC in 2005 evolved into the research project Mathematics Problem Solving for Everyone, or MProSE, in 2008. Thus, the MProSE way of teaching problem solving has been one of the key features of our IP mathematics curriculum since 2005, albeit at different levels of intensity over the years. It also varied slightly in approaches and assessment mode

---

[5] New teachers refer to teachers who had not gone through the same training conducted by the research team.

from year to year. Over the past eight years, different teaching pedagogies and innovations (e.g., crafting ICT-based problems) have been incorporated into the programme to equip our students with necessary skills and techniques to solve mathematical problems. In the following paragraphs, we will describe and discuss some teaching approaches and assessment practices that we have implemented for the problem solving module in the past few years.

At the beginning of the first semester, all classes of Year 9 students went through a mathematical problem solving module using Pólya's four-stage problem solving model. Schoenfeld's problem solving framework was also introduced to the students with the objective of getting students to understand the various factors that can influence their problem solving process. During the module, students learnt to use different problem solving heuristics through different types of problems. They were also made aware of some of the unproductive beliefs that they may have about problem solving and how these beliefs can influence their ability to solve a problem. Through understanding these factors, it was hoped that students' learning attitudes would improve and that students would become effective problem solvers with good meta-cognitive skills. During lessons, students learnt collaboratively in groups which encouraged and supported the practice of exchanging ideas when they were given challenging mathematics problem. Students were also taught to reflect on their problem solving steps using the Practical Worksheet which was specially designed to provide greater scaffolding as students worked through the four stages of problem solving.

Besides conducting the mathematical problem solving module at the beginning of the first semester at the Year 9 level, many opportunities were also provided for students to use their problem solving skills in the various mathematics topics taught throughout their four years of study at TJC. These were carried out through class discussion, homework, and continuous assessments.

Students' problem solving skills were assessed at their mid-year and year-end examination. At each examination, students sat for two papers. In Year 9, Paper 2 of both the mid-year and the year-end examinations are designed to assess students' problem solving skills. In Year 10, there will also be one question in each paper that tests students' problem solving skills.

Our students use technology extensively during their mathematics lessons. In particular, they work with graphing technologies through mathematics software such as TI-Nspire software. The software provides a multi-representative approach that helps students have a better diagrammatic visualisation of problems which should lead to a better understanding of the problems (Kaput,1992; Burrill et al., 2002). With the availability of the support of visual representations by the software, our assessment tasks have also evolved into tasks that involve more dynamic visual displays. These assessment tasks were also designed to support different forms of mathematical processing and understanding (Lowrie, 2012). Furthermore, non-routine problems that required the use of specific problem solving heuristics can also be designed with greater flexibility by the teachers.

Hence, students were allowed to use the software in Paper 2 of their examination to help them arrive at solutions that would otherwise have been too time-consuming to come up with due to the tedious computations required. In some cases, these problems, combined with the use of the software, provided opportunities for students to arrive at the same solution using more than one method. From the alternative methods, they were able either to see the connections so as to derive new knowledge or do a check of their analytical solutions or conjectures.

4 Some Selected Mathematics Problems

In this section, we share some of the problems we used for the problem solving module and provide a rationale for including these problems. We also share some of the problems we posed during the examinations to illustrate how we integrated ICT in these problems.

These problems required only knowledge at the lower secondary mathematics levels. These problems also provided opportunities for students to practise the use of Pólya's four stages in problem solving.

Example 1. (A problem used in class work) The figure below shows one of the problems used in one of the classroom sessions.

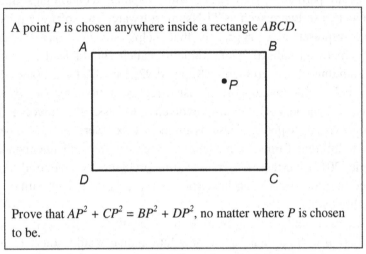

A point P is chosen anywhere inside a rectangle $ABCD$.

Prove that $AP^2 + CP^2 = BP^2 + DP^2$, no matter where P is chosen to be.

Figure 5. A classroom problem for mathematical problem solving

Rationale for introducing this problem as a classroom exercise. We introduced this problem as a class exercise because we wanted students to be aware of and to use certain problem solving heuristics that we thought were important for students to know. In particular, we wanted students to consider a special case in their attempts to solve the problem and draw suitable diagrams.

In this problem, we thought that students could consider the case where P is chosen to be at the centre of the rectangle. For this special case, obviously AP = CP = BP = DP and thus the result is trivially true. From this special case, students can then perhaps gain more insight into solving the more general case. In particular, students may find that drawing auxiliary lines from the vertices of the rectangle to P and drawing the horizontal and vertical lines passing through point P may be productive ways to proceed. Figure 6 provides a diagram of how we imagine students can make use of drawing auxiliary lines for this problem. From this diagram, perhaps students can then access their prior knowledge of the Pythagorean Theorem to solve the problem. In essence, we wanted students to realise how each of the terms in the desired result,

$AP^2 + CP^2 = BP^2 + DP^2$, is the square of the length of a hypotenuse of a right triangle in Figure 6. Thus, application of the Pythagorean Theorem and some algebraic manipulation can then lead them to proving the equation.

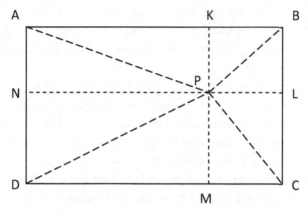

Figure 6. Important supplementary lines drawn for the problem in Example 1

Example 2. (A homework assignment problem) Though rate and speed are familiar topics for our students, the nature of this topic provided teachers with a lot of flexibility and creativity to set interesting and unfamiliar problems. The problem in Figure 7 was given as a homework assignment for a Year 9 class.

Two cyclists, A and B, are 30 kilometres apart and are travelling directly towards each other, in a straight line. Cyclist A is moving at a constant speed of 8 km/h while cyclist B is moving at a constant speed of 12 km/h. At the same time, a bee starts from A and flies towards B in a straight line at a constant speed of 35km/h. When it reaches B, it immediately turns around and heads back towards A in a straight line, without slowing down. It will continually fly back and forth between A and B until they meet. How far will the bee have travelled by then?

Figure 7. A homework problem

Rationale for introducing this problem as homework. This problem was given to students as homework so that students would have more time to work on the problem independently using the Practical Worksheet. At the time this homework was given, students were already introduced to using various problem solving heuristics. The key learning objective for this problem was for students to practise the first two stages of Pólya's four-stage problem solving model (i.e., Understand the problem and Devise a plan).

Students were required to make use of the given information and their knowledge of the relationship between distance, speed and time to devise a plan. One possible step that could have been taken was through working backwards. Since distance travelled by the bee was the required aim of the question and the speed of the travelling bee was given, the problem effectively required students to first find the time taken for A and B to meet. Thus a possible solution can be as follows:

> Cyclists A and B meet when the sum of distances travelled by cyclists A and B is 30 km.
>
> Let the time required for A and B to meet from the time they start moving be x hrs.
>
> Thus,
>> distance travelled by A = $8x$ km, and
>> distance travelled by B = $12x$ km.
>
> Total distance travelled by A and B = $8x + 12x = 20x$ km.
> Hence, the time it took for A and B to meet, $x = 30/20 = 1.5$ hrs.
> Thus, distance travelled by bee = $(1.5)(35) = 52.5$ km.

Example 3. (Another homework assignment problem) Algebra is an important mathematics topic for our IP students. Thus, the mathematical competency of using algebra in problem solving or manipulating algebraic expressions and equations is a fundamental skill that is required at higher levels of our mathematics curriculum. A problem we posed as homework involving more algebraic manipulation is given in Figure 8 below.

> The sum of two numbers is 5 and the product of the two numbers is -8. Find the exact sum of the squares of the reciprocals of the numbers.

Figure 8. Another homework problem

Rationale of introducing this problem. The problem aimed to make students aware that an algebraic expression (no matter how complicated it may be) can be seen as one "entity" in some problem situations. In this respect, students were required to, firstly, use the algebraic representation of "squares of reciprocals of the numbers" to help them identify the steps needed in this problem, and, secondly, see the value of the "squares of reciprocals of the numbers" as one mathematical object that can be obtained without finding its constituents. In the process of solving the problem, they were required to make appropriate connections to their prior knowledge in algebra such as completing the square in order to solve the problem.

Example 4. (A problem in the Year 9 mid-year examination) Students had not been taught the topic of permutation and combination. Hence the question found in Figure 9 which was posed during the Year 9 mid-year examination came across as a real life non-routine problem for the students. They needed to use systematic listing as the heuristic to devise a plan for solving the problem.

William is going on a road trip in the morning and wishes to get some socks from his drawer. Not wishing to wake his wife and son, he resorts to getting socks from the drawer in the dark. Suppose that he has a large number of socks in the drawer and all of them are identical in size and shape. However, they come in 7 different colours. Find the minimum number of socks he has to take out of the drawer to ensure that he has at least 5 matching pairs.

Figure 9. A problem in the Year 9 mid-year examination

Example 5. (A problem in the Year 9 year-end examination)

Figure 10 shows a problem in the Year 9 year-end examination. This is an example of a problem that lent itself to the use of ICT devices. This problem started with a simpler problem by getting students to solve the equation

$$\sin x + \sin 3x = 0, \text{ where } 0° \leq x \leq 180°.$$

This was followed by the use of systematic listing to observe patterns in the number of solutions to the equation $\sin x + \sin kx = 0$, where $0° < x < 180°$. The last stage involved assessing students' ability to generalise from the observed pattern and come up with the number of solutions to the equation $\sin x + \sin nx = 0$ for a general positive integer n, where $0° \leq x \leq 180°$.

1. Using algebraic methods *only*, find the solutions to the equation

$\sin x + \sin 3x = 0$ where $0° \leq x \leq 180°$.

2. The given graph shows $y = \sin x + \sin kx$, where k is a value controlled by a slider. The value of k is an integer that ranges from 1 to 20.

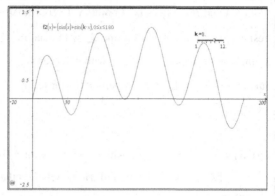

By varying the value of k, observe the number of distinct solutions to the equation
$\sin x + \sin kx = 0$ where $0° \leq x \leq 180°$.

(There is no need to determine what the solutions are.)

Complete the table below.

Value of k	Number of Distinct Solutions	Value of k	Number of Distinct Solutions
1	2	6	
2	3	7	
3		8	
4		9	
5		10	

3. (a) Write down the number of solutions to the equation $\sin x + \sin 99 \, x = 0$, where $0° \leq x \leq 180°$.

(b) Write down the number of solutions to the equation $\sin x + \sin 100 \, x = 0$, where $0° \leq x \leq 180°$.

(c) Write down the number of solutions to the equation $\sin x + \sin n \, x = 0$ for a general positive integer n, where $0° \leq x \leq 180°$.

Figure 10. A problem in the Year 9 year-end examination

Example 6. (A problem in the Year 9 mid-year examination)
Another problem in a Year 9 examination is shown in Figure 11. This problem also required the students to use ICT devices.

Points A and B are two fixed points with coordinates $(-2,1)$ and $(1,2)$ respectively. Point A lies on line L_1 with equation $kx + y - 1 + 2k = 0$ and point B lies on another line L_2 with equation $ky - x + 1 - 2k = 0$ where k is any real number. Line L_1 is perpendicular to line L_2 and they intersect at point P.

(a) (i) By rotating line L_1 while keeping point A fixed, find the **integer** value of k for which the area of triangle APB is the greatest.

   (ii) Points C and D are points on line L_1 and L_2 respectively such that P is the midpoint of AC and also midpoint of BD. Explain why quadrilateral $ABCD$ is a square when the area of triangle APB is the greatest.

   (iii) Another point Q is such that the area of triangle AQB is equal to the greatest area of triangle APB. State the equation of a straight line on which the point Q lies.

(b) Using geometry, trace and observe the shape of the path traced out by moving point P (x, y) as k varies. Construct the shape of the path that passes through the points traced out by point P.

   (i) Use the software to estimate the equation of this path in the form:
   $$(x-a)^2 + (y-b)^2 = r^2$$
   where a, b, and r are positive real constants to be determined.

   (ii) Point M is the midpoint of line segment AB. Find the **exact** coordinates of point M and the **exact** length of AM.

   (iii) By comparing (b)(i) and (b)(ii), describe the significance of a, b and r in relation to the shape of the path traced out by point P.

Figure 11. Another problem in the Year 9 mid-year examination

This problem was a non-routine question where students needed to see connections via two different approaches – coordinate geometry and visualising dynamic changes using ICT – to arrive at understanding the equation of a circle. Students have not learnt the equation of a circle before this question was posed to them and thus this was a challenging task for them.

If done correctly, students will arrive at a figure similar to the one shown in Figure 12. Students were assessed based on their ability to make the connection between and draw out the significance of the coordinates of the centre of the circle and the length of the circle's radius to the equation of the circle $(x-a)^2 + (y-b)^2 = r^2$.

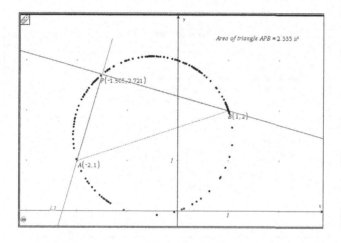

Figure 12. A possible graphical display output of the problem in Example 6

5 Some Comments on Students' Works on the Worksheet

Through students' use of the Practical Worksheet, we were able to assess students' proficiency with the problem as well as their appreciation of the problem solving process. We illustrate using extracts of a certain student's work on the speed problem detailed in Example 2 in the previous section. The student's working is given in Figures 13 to 16.

In the first stage of understanding the problem (Figure 13), the student identified 'drawing a diagram' as a heuristic to help her. Particularly, she made use of a line model that indicated the different conditions in the problem. More accurately, she wrote that she drew a diagram "to show the relationship among all the subjects".

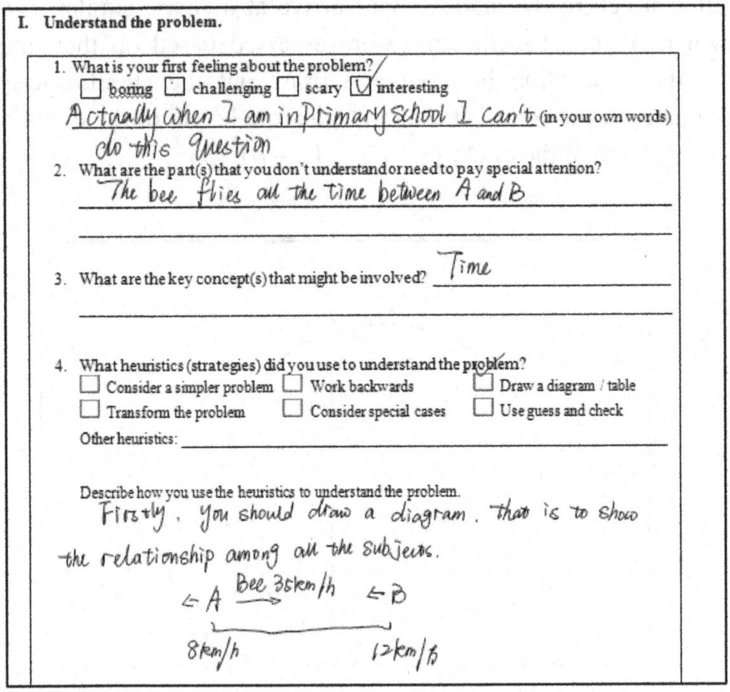

I. Understand the problem.

1. What is your first feeling about the problem?
   ☐ boring ☐ challenging ☐ scary ☑ interesting
   Actually when I am in Primary School I can't (in your own words)
   do this question

2. What are the part(s) that you don't understand or need to pay special attention?
   The bee flies all the time between A and B

3. What are the key concept(s) that might be involved? *Time*

4. What heuristics (strategies) did you use to understand the problem?
   ☐ Consider a simpler problem   ☐ Work backwards   ☐ Draw a diagram / table
   ☐ Transform the problem   ☐ Consider special cases ☐ Use guess and check
   Other heuristics: _____

   Describe how you use the heuristics to understand the problem.
   Firstly, you should draw a diagram. that is to show the relationship among all the subjects.

   ←A Bee 35km/h → ←B
   8km/h   12km/h

Figure 13. A student's response to Stage 1 (Understanding the problem)

In Stage 2 (Figure 14), the student wrote down her plan for solving the problem. Note that she once again mentioned drawing a diagram and the relationship between the different subjects. Clearly, this student correctly thought that a key to the problem's solution was determining a relationship between the two cyclists and the bee. She eventually arrived at an incorrect solution, however, and her written work under Stage 3 (Figure 15) revealed that there were some gaps in her understanding of the mathematical concept behind the problem.

II. Devise a plan.
Write out the steps of your plan briefly in point form on how you are going to solve the problem.

Plan 1:

1. draw the diagram to show the things clearly

2. to think about the relationship among all the subjects

3. to find the same points of all the subjects

4. Solve it.

Figure 14. A student's response to Stage 2 (Devise a plan)

In solving for the distance travelled by the bee, the student clearly knew that she would have to use the formula which equated distance to the speed multiplied with time. And since she already knew that the bee travelled at the speed of 35 km/h, this meant that she just needed to find the time it took the bee to fly. Her solution indicated that she recognised that this required time was just equivalent to the time it took for cyclist A and B to meet. She was correct in identifying this relationship. However, she computed for the time it took for the cyclists to meet erroneously. She divided the distance of 30 km by the difference of the speeds of cyclist A and B. The student gave no explanation for this computation, but it seemed that by obtaining the difference of the speeds of the two cyclists, she may have erroneously applied the concept of "resultant speed".

In the fourth stage (Figure 16), the student does not actually conduct a check of her solution. She offered an alternative method of using equations to look for the required time but did not carry it out. Finally, she proposed an extension by modifying the problem such that cyclist A and B travel in different directions. While this created an interesting situation, the problem itself would not make sense since the question was dependent on the moment when the two cyclists eventually meet. It would seem that the student still had a vague comprehension of what Stage 4 required from her.

In general, we found that majority of students had difficulty in Stage 4. Some students thought of simply repeating the calculation as a form of checking. Most students did not provide adaptations, extensions or generalisations.

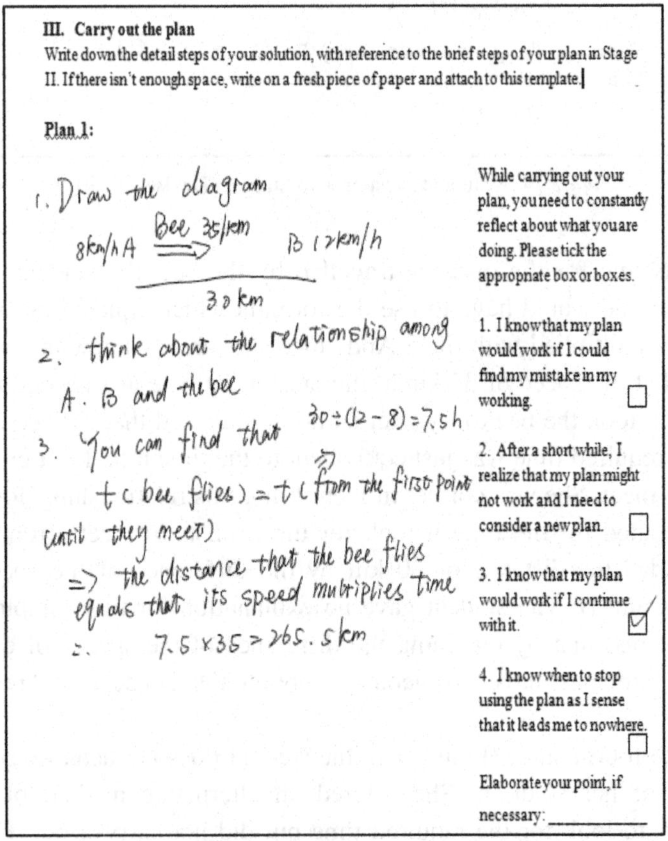

III. Carry out the plan

Write down the detail steps of your solution, with reference to the brief steps of your plan in Stage II. If there isn't enough space, write on a fresh piece of paper and attach to this template.

Plan 1:

1. Draw the diagram

8km/h A —Bee 35/km→ B 12km/h

30 km

2. think about the relationship among A. B and the bee

3. You can find that $30 \div (12-8) = 7.5h$

t (bee flies) = t (from the first point until they meet)

=> the distance that the bee flies equals that its speed multiplies time

∴ $7.5 \times 35 = 265.5km$

While carrying out your plan, you need to constantly reflect about what you are doing. Please tick the appropriate box or boxes.

1. I know that my plan would work if I could find my mistake in my working. ☐

2. After a short while, I realize that my plan might not work and I need to consider a new plan. ☐

3. I know that my plan would work if I continue with it. ☑

4. I know when to stop using the plan as I sense that it leads me to nowhere. ☐

Elaborate your point, if necessary: ________

Figure 15. A student's response to Stage 3 (Carry out the plan)

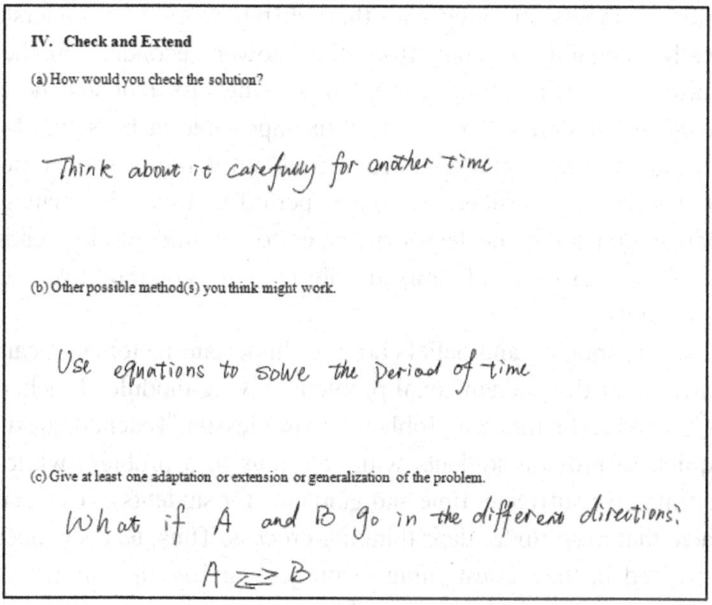

IV. **Check and Extend**

(a) How would you check the solution?

Think about it carefully for another time

(b) Other possible method(s) you think might work.

Use equations to solve the period of time

(c) Give at least one adaptation or extension or generalization of the problem.

What if A and B go in the different directions?

A ⇄ B

Figure 16. Student's response to Stage 4 (Check and expand)

6 Challenges and Looking Ahead

There were several challenges that teachers faced in the implementation of mathematical problem solving in our curriculum. One of the key stages of Pólya's problem solving model includes the stage of expanding the problem. However, observing from students' sample worksheets, even though students were taught how to use the worksheet and had experienced using the worksheet in class, they were unable to effectively attempt this stage independently when these problems were issued as homework assignments. Thus, teachers needed to spend some time in class discussing this stage even if these problems were given out as homework assignments.

Our Year 9 students came from different backgrounds[6] and

---

[6] This was the case when the IP in TJC was in the transitory phase where it did not yet extend to Year 7 and 8 levels.

secondary schools, bringing with them different levels of understanding of mathematical knowledge from their lower secondary mathematics education. As such, setting a problem solving task that will be equally received by all students was almost an impossible task. Naturally, there were students who could complete a problem within a short time while others required a considerably longer period of time. We managed the challenge of pacing the lesson to cater to all students by seating the students in groups of mixed ability in collaborative learning environments.

Teacher training and beliefs are also important factors that can affect the success of the mathematical problem solving module. Teachers need to be trained to facilitate a problem solving lesson. Teachers must not be too quick to provide students with solutions to a problem while at the same time give sufficient time and guidance for students to overcome any obstacle that may hinder their thinking process. Thus, teachers need to be well-trained in their questioning techniques and getting students to use the problem solving template so that students can move towards self-regulating their problem solving process.

From an attitudinal survey conducted in 2012 with 30 Year 9 students before and after they were introduced to the problem solving module, the combined percentage of students who strongly agreed or agreed with the statements, "I am confident with applying problem solving skills" and "I get a sense of satisfaction when I solve mathematics problems" were obtained and summarised in Table 1. The students maintained a high sense of satisfaction with the task of solving mathematics problems. As to confidence in applying relevant problem solving skills, there was an increase of ten percentage points.

Table 1
Attitudinal survey result of Year 9 students on mathematical problem solving

Statement	Pre-survey	Post-survey
I am confident with applying problem solving skills.	70%	80%
I get a sense of satisfaction when I solve mathematics problems.	96.7%	96.7%

Starting from 2013, TJC's Integrated Programme expanded to encompass six years as we began to take in Year 7 students. As part of our continual effort to achieving our aim of providing students with a more systematic approach to solving non-routine problems as well as to stretch and add value to students' intellectual development in mathematics, the problem solving module will be introduced to our first batch of Year 7 students. With the growing emphasis and importance of introducing non-routine problems to students in their early secondary school years (Yeo, 2009), the proposed plan will include introducing these problems in the first semester and focusing on introducing ideas of problem solving skills in topics on numbers, algebra and geometry. This will be followed by a formal introduction of the new MProSE problem solving module in the second semester.

Looking ahead, even though there are challenges that we, as teachers, need to overcome, we strongly believe that mathematical problem solving plays an important role in shaping our students' minds by getting them to construct and assess their own mathematical understanding and thinking.

Chapter 3

Mathematical Problem Solving in NUS High School

Grace TAN Kok Eng   Joyce SEOW Chwee Loon   TAN Boon Keong
TOH Tin Lam (NIE Researcher)

In line with the articulated goal of the Singapore mathematics curriculum, the National University of Singapore High School believes in the importance of developing and strengthening our students' problem solving abilities. We thus collaborated with a team of researchers from NIE to help us in developing and implementing a curriculum that could better achieve this. In this chapter, we share our experience of participating in this collaboration and our reflections as a result of this participation.

1 Introduction

Mathematical problem solving is the heart of the Singapore mathematics curriculum for all its students from the primary to the pre-university level regardless of ability. It is thus important in Singapore's mathematics education for all mathematics teachers to equip our students with the ability to do problem solving. In particular, this is the belief of the Mathematics Department of the National University of Singapore (NUS) High School which is a school that specialises in mathematics and science.

In 2008, NUS High School's Mathematics Department worked with a team of researchers from the National Institute of Education (NIE) to work on incorporating problem solving into our curriculum, through the research project Mathematical Problem Solving for Everyone (MProSE). According to MProSE, problem solving is the core of mathematics education, and students should be made more explicitly aware of the processes of problem solving instead of merely focusing on the correctness of the solutions of mathematics problems. As the NUS High School Mathematics Department held the same beliefs about the learning of mathematics, we proceeded to work with the team from NIE to incorporate problem solving into our mathematics curriculum. We believed that we can better and more meaningfully engage students in mathematical thinking through the problem solving model proposed by MProSE.

2 MProSE Implementation

Professional Development. The initial involvement of the Mathematics Department of NUS High School with MProSE began with the professional development programme for all mathematics teachers in NUS High School. All the teachers in the department underwent a 12-hour training with NIE on teaching problem solving. During the training sessions, the instructor, being one of the members of the MProSE team, played the dual role of modelling how problem solving could be taught in a secondary school classroom and explicating the pedagogy of teaching problem solving, including the introduction of Pólya's model. In the series of workshops, the participants also had the opportunity to seek the NIE researchers' help in solving potential issues that might arise during the implementation of this new initiative. Many of the problems used during the professional development were ultimately used as the set of MProSE problems.

Initial Implementation. MProSE, being a design experiment, worked on some parameters. If the school were to embark on MProSE, (1) assessment must be an essential component, and (2) the module should be made compulsory for all students and not be offered merely as an elective. The first round of MProSE took place in the first semester of 2009. As it was still at an experimental stage and teachers were not familiar with how to carry out the teaching of the relatively new approach of problem solving, the NIE researchers agreed that the MProSE module be first introduced in NUS High School as an elective module. This pilot trial was taught by Tay Eng Guan from the NIE research team. The teachers had the opportunity to observe the lessons within a Lesson Study context. Following each lesson, a meeting between the teachers and the NIE research team was conducted to address the concerns of the teachers and identify ways to fine-tune the lessons so that the teachers would be comfortable with the programme before they taught it. The MProSE module was then introduced to all Year 8 students of the school.

Full Implementation. Our teachers took over the teaching of MProSE lessons in the second semester of 2009 when the MProSE module was first incorporated into the core mathematics curriculum of the school for all Year 8 students. This module was taught by the first group of teachers who had attended the MProSE professional development workshop conducted by the NIE researchers. The NIE research team provided full support for this round of teaching by observing the teachers' lessons and providing constructive feedback to the teachers after each lesson. Videos of the teaching and class responses were also recorded. These records served as very good feedback for the teachers' self-improvement. Subsequently, more teachers embarked on teaching the MProSE module. The collection of videos also served as resources for the latter groups of teachers teaching the MProSE module.

In the second semester of 2010, another new group of teachers taught the MProSE module in the core curriculum. This group of teachers had several discussion sessions with the previous group of teachers who taught it in 2009. During these discussions, strategies on the implementation of the lessons were shared. Potential problems and assessment structures were also discussed during these sessions. These discussion sessions proved to be very fruitful and useful as they provided the second group of teachers more confidence to implement the MProSE module with the students. The teachers also had the opportunity to watch the videos recorded by the previous batch and were better prepared when they went into the classes to teach.

Using this form of implementation where teaching teams build on the experience of the earlier teaching teams, more teachers were trained over time, resulting in more teachers becoming confident in teaching and carrying out problem solving in the classroom. The discussion sessions also allowed peer sharing and collaboration which strengthened the ties within the department. Over time, a community of practice was established.

During the initial teacher professional development provided by the NIE researchers and the subsequent professional sharing among teachers in the Mathematics Department, the pedagogical skills of the mathematics teachers were developed. In particular, they were able to see how Pólya's four stages could be employed to solve mathematics problems.

Structure of the Problem Solving Module. MProSE was introduced to the Year 8 students in the core mathematics module of the second semester. The lessons spanned ten hours in the curriculum, including the final session where a test[1] using the Practical Worksheet was given. (See Appendix A.1 for the Practical Worksheet.)

During the first three years of implementation, the teachers followed the NIE-recommended lesson plans closely. There were not many changes until the later years in 2012, when the teachers became more confident in their teaching and gained more clarity regarding the objectives of problem solving.

Following the NIE prescribed lesson plans, Pólya's problem solving stages were introduced to the students during the first three lessons. After the introduction of the stages, students were encouraged to write these stages on butcher sheets which were placed around the classroom to remind students of the stages while they worked on the Practical Worksheets for the other five lessons. This was a conscious attempt made by the teachers to reinforce the problem solving stages in class.

For the first lesson, students were introduced to the first two stages of Pólya's problem solving model, i.e., Understanding the Problem and

---

[1] Such a test will be subsequently referred to as the Practical Test for the rest of this book.

Devising a Plan. They were required to practise these stages with the various problems provided. The key in the lesson was not for the students to solve the problems but rather, to understand the problem and devise a plan to solve it. The lessons were carried out interactively such that students were given the opportunity to devise various plans and share these in class. Very often, there was more than one approach to solve a particular problem and students were excited over the different strategies that they had come up with.

For the second lesson, students were taught the third stage of Pólya's problem solving model, i.e. Carrying out the Plan. They were given a problem that gave them an opportunity to practice the first three stages. Time was allocated for students to present their plans in class where they learnt from one another the various strategies that could be used to solve the problem.

The last stage, Check and Expand was taught in the third lesson. Students were required to look for alternative solutions and were asked to attempt to generalise and extend the problems given in the earlier two lessons. Class time was also allocated for students to present their generalisations and extensions in class.

In the first three lessons, we allow students to work in groups. We think this will help students – especially those who started off with less confidence in problem solving – develop the right affective disposition to ease into the module.

After the first three lessons, students were ready to solve problems using Pólya's four-stage model. For each subsequent lesson, a problem printed on a Practical Worksheet, was given to students. They were required to work through the first two stages in class. Students who had difficulties in the first two stages were given more assistance and hints by the teachers. Once they had finished the first two stages, students were encouraged to solve the problem, look for alternative solutions and further expand the problem. These remaining two stages were given as homework. Students would then submit their work the next day. The teacher would mark their work and go through the problem with the students in the subsequent lessons.

In 2012, MProSE was taught via the 'flipped classroom' approach. A certain problem was chosen and solved by the teacher using Pólya's problem solving model. This lesson was recorded and uploaded on the school's server. Students were given the notes and solutions to the problem and they were expected to watch the videos and understand Pólya's problem solving model before the lesson.

For the first lesson, the teacher recapitulated Pólya's stages with the students. In groups, students practised the first two stages using four specifically chosen problems reproduced in Figure 1 below. This took about half an hour and the remaining time was given for the students to present their strategies for solving the problems.

---

1. Find the last digit of 2009^{2009}.

2. A boy claims that when left class A and joined class B, he raised the mean Math scores of both classes. Explain if this is possible.

3. Two squares, each s on a side, are placed such that the corner of one square lies on the centre of the other. Describe, in terms of s, the range of possible areas representing the intersections of the two squares.

4. A 'nice' number is a number that can be expressed as the sum of a string of two or more consecutive positive integers. Determine which of the numbers from 50 to 70 inclusive are nice.

Figure 1. The four problems used to introduce Pólya's stages in 2012

The same four problems were used for the second lesson. This time, students were required to apply Pólya's last two stages to solve these problems. The same four problems were used to allow students to concentrate on applying Pólya's stages so that they will not be distracted by having to tackle completely different problems. This also helped students to reinforce the stages.

Students then presented their solutions and the various strategies and extensions in class during the third lesson.

More classroom time was 'freed up' via the flipped approach. The extra class time was used for discussions and discourse in class. Students learnt from their peers, and teachers acted as facilitators to stretch students' potential through questions such as "What if we change this (condition)?", "Why are we able to use this (method)?", "Is this solution (strategy) applicable to all problems?" and "What are some of the restrictions?" This method of teaching encouraged students to think about and understand the concepts more deeply.

Assessment Structure. We adopted the MProSE rubric (see Appendix C) to assess the students' answers on the Practical Worksheets. We find the MProSE rubric comprehensive and extensive – it is easy to understand and use. In using the rubric, we were aware that the main emphasis was in identifying students' work that evidenced the use of Pólya's stages. They were also graded on their ability to apply the heuristics and come up with alternative solutions to the problems. Students were awarded more marks if they were able to extend their problems further.

In NUS High, the MProSE module constituted 9% of the overall core module grade. Students' grades for the MProSE component were based solely on a Practical Test after the end of nine lessons. They were required to solve the given problem in an hour using Pólya's four-stage model.

However, this assessment structure might not have assessed students' overall problem solving ability because it was based on only one Practical Test. It was proposed that future cycles include components such as class participation, where other aspects of the students' learning processes are also graded, in addition to the summative Practical Test which assesses the product of their learning. In this way, we may better reinforce the students' appreciation of the importance of mathematical problem solving.

Hence, in 2012, we tried a different assessment structure. The problem solving module was assessed based on various components: (1) the students' ability to apply the stages as observed during the first three lessons, (2) an average score of students' best two works on the Practical Worksheets, and (3) the Practical Test.

3 Reflections

The following observations were made by the teachers teaching MProSE lessons. At the beginning stage, the students were resistant to using Pólya's four-stage model in solving their mathematics problems. They were more focussed on obtaining the correct answers to the given problems rather than the problem solving process itself. However, with constant encouragement and guidance from the teachers, the students slowly got into the habit of following Pólya's four stages. The structure of the Practical Worksheet and the assessment rubrics also ensured that the students went through the process of Pólya's four-stage model. Gradually, the students were tuned to this model and were able to complete their Practical Worksheets with few reminders to utilise the stages.

The students we deemed as demonstrating higher ability in mathematics generally reacted positively to the challenge of new problems and they were also stimulated by the prompts offered by teachers to find alternative solutions or generalise the problems. For these students, the teacher also needed to prepare more problems on hand so as to further stretch their abilities. Hence, teachers teaching problem solving modules needed to be highly competent in mathematics content knowledge in order to react to the students' questions.

Another challenge that the teachers faced was the fear of coming across a situation wherein a student would be asking a question in class which the teacher might not be able to provide an answer for. However, in the MProSE module, the teachers acted more as facilitators rather than providers of solutions. Instead of giving direct solutions to the students, teachers focussed on engaging the students in the thinking process and working through the problem with them. There had to be a change in the mindset on the role of the teacher from being a provider of knowledge to a facilitator of learning. In this way, learning took place through discourse and discovery.

At NUS High School, our students had opportunities to participate in class discussions about the range of approaches used and the range of solutions found during the problem solving process. Together with thoughtful questioning by the teachers, our students achieved new levels of engagement and learning. Through the class discussions, teachers also learned from the students.

Teachers teaching MProSE also met up regularly to discuss teaching strategies and share different approaches and solutions to the problems discussed. Teachers also attended workshops related to problem solving to build their capacity for problem solving and it teaching.

Students who were consistently in the lower performance bands in mathematics needed closer attention, guidance and encouragement from the teachers but they were able to participate in class discussions.

Another challenge we faced was the need to choose appropriate problems which would incorporate mathematics concepts that the students had learnt in the regular curriculum. The problem of course had to be solvable, but it ideally also needed to have alternative solutions, be extendable and generalisable, and be illustrative of the problem solving processes. Teachers had difficulties looking for more of such problems. The book *Making Mathematics Practical: An Approach to Problem Solving* (Toh, Quek, Leong, Dindyal, & Tay, 2011) was the main reference to the problems suitable for problem solving. Our teachers also looked for more mathematics problems on the Internet and from the collection of students' work.

4 Conclusion

Problem solving was successfully implemented in NUS High School due to the strong support given by the department and school leaders. The school provided extra time in the curriculum for the problem solving module. Our principal, Hang Kim Hoo, also provided his valuable advice on the implementation of this module. Without their support and encouragement, it would not have been possible for the team to work through the challenges of this new initiative. Teamwork among the teachers was also vital to the development of teaching materials and teachers' capacity for the teaching of problem solving.

Through the problem solving approach, our students have been trained to be adaptive learners so that they can perform effectively when faced with novel problems in the future. Pólya's four-stage model helps them to regulate their problem solving attempt. In the long run, we hope the students will internalise these habits of mind.

In addition, during the problem solving process, developing control was necessary while looping through Pólya's four stages. This helped train our students to become aware of and reflect on their thinking processes. This metacognitive ability serves as a building block to develop our students to become independent and self-directed learners.

Finally, in the next phase, we hope to raise the status of problem solving to become an integral part of mathematics learning by moving beyond teaching problem solving explicitly to teaching via problem solving.

Comments from the NIE Researcher.

NUS High School was the first school to work with the NIE research team where we officially embarked on MProSE as a funded research project. The MProSE project involved a radical shift of the emphasis from the usual school mathematics curriculum to the processes involved in problem solving. Not only did the researchers complete one full cycle of research with the school, the MProSE problem solving module has now been made a compulsory module for all students in this school.

It should also be noted that the MProSE design was not adopted wholesale from the researchers even in the initial phase; it was adapted to meet the school's needs and the constraints of reality. The researchers' original intention was to have the MProSE problem solving lessons as the "practical" lessons for the school's usual mathematics lessons. However, due to the modular structure of the courses of the school, the MProSE problem solving module was developed instead of the practical lessons that run parallel to their core mathematics lessons.

In this chapter, the teachers from NUS High School reported their observations, the successes, and challenges they encountered in teaching problem solving. It is the hope of the NIE researchers that the points described by the authors of this chapter will provide useful insight to readers who might be interested to implement such a problem solving curriculum in their schools.

NUS High School is a school that specialises in mathematics and science so it might be asked how the MProSE module can be adapted and used in other schools. The researchers' intention in starting with

NUS High School was to develop a "success case", which would enable us subsequently to tweak the research to better fit the structure of the mainstream schools. In fact, the subsequent chapters of this book are reported by the other mainstream schools which have attempted to carry out MProSE in their schools.

Chapter 4

Mathematical Problem Solving in Tanjong Katong Girls' School

CHEONG Chui Chui Hilary   Grace LEE Yun Yun

TAY Eng Guan (NIE Researcher)

This chapter aims to consolidate the experience of Tanjong Katong Girls' School in teaching problem solving using the MProSE programme to eight Year 8 classes over nine lessons. We first share our motivation for participating in the programme. We then review the students' test scores, students' feedback, and teachers' reflections to uncover the outcomes of our first implementation. In general, we found that despite certain challenges and setbacks, MProSE can benefit all our students as it can prepare them for the 21st century by equipping them with essential problem solving skills. Thus, in the last part of this chapter, we describe the improvements we made for our second implementation to better achieve the intended results of teaching problem solving to our students.

1 Background

The 21st century knowledge-based society calls for mathematics education to train learners to be flexible, creative, confident and good team players who are able to solve new problems and deal with ambiguities; the most efficient workers who can only follow standard procedures and ace examinations and tests may not survive in the world tomorrow (Tan, S.C., Divaharan, Tan, L., & Cheah, 2011). As such, we at Tanjong Katong Girls' School (TKGS) desire to nurture creative and confident problem solving students for the 21st century. However, we find that it is not easy to carry this out.

When we were introduced to the Mathematical Problem Solving for Everyone (MProSE) programme, we thought that it could help us prepare our students for the 21st century by offering much more than what was promised in theory. Its approach was strongly aligned with our school's holistic goals of character development and nurturing creative problem solvers and self-directed learners, preparing students to better face challenging problems in life and serving the community through problem solving. It established a common mathematical language for problem solving essential for quality teaching and learning across the levels. MProSE focus also on the process, not just the product.

As such, we felt that MProSE could raise the quality of math education across levels and abilities. The high ability learners could be stretched to think out of the box during problem solving, especially during the check and expand step. The middle ability learners could gain much by learning Polya's stages and heuristics. We also thought that it could address the needs of students who were more academically challenged by mathematics and those who have developed an aversion towards learning mathematics as MProSE give students firstly the space to communicate their feelings and address their fears and secondly, the structure that guides their thinking habits.

Because of the benefits that we believed our students could receive from the design, our school adopted MProSE's approach to teaching problem solving as part of the school's 4-year mathematics programme. We would like to add, however, that we were also strongly motivated to undertake the programme because of the professional development support that we were assured to receive from a team of committed university mathematics professors from the National Institute of Education (NIE) who would be working on the ground with us. This support was crucial. Left on our own, we might not be able to implement the programme properly, sufficiently face the challenges ahead, and achieve the results we desired. We certainly valued the depth of knowledge that the NIE researchers contributed to our teachers' professional development as they modelled the problem solving process and its instruction.

We started planning for implementing MProSE in our school in August 2011. It was implemented in January 2012 for the cohort of Year 8 learners. We found that MProSE was able to address the gaps within our mathematics curriculum to develop creative problem solvers and it made successful problem solving within reach for many learners. We share our story in the succeeding pages.

2 First Implementation of MProSE in TKGS

We wanted to introduce MProSE to students as early as possible. However, we felt that Year 7 students would require some time to adjust to secondary school life. In relation to this, the mathematics department's focus for Year 7 students was bridging their mathematics learning and problem solving skills from primary school to that which was demanded in secondary school. The explicit teaching of MProSE was carried out with our Year 8 students. At this level, we believe that students would have already developed the discipline and maturity that the MProSE programme required. Furthermore, students would already possess the foundation for algebra which would be useful in dealing with some of the problems.

All 280 Year 8 students attended the nine 1-hour MProSE lessons conducted after school by the mathematics department teachers spread out over nine weeks. In the course of the programme, the students were introduced to the four key stages of Pólya's (1945) problem solving strategy which involved understanding the problem, devising a plan, carrying out the plan, and checking and expanding. The students were presented with various Problems of the Day (PoD) which are open-ended and context-based yet unconventional in nature.

For each PoD, students were required to go through all four stages of Pólya's problem solving strategy by filling up the Practical Worksheet. In doing so, students were made to recall and recognise the different bodies of mathematical knowledge they would require. They also penned down the mathematical heuristics and processes necessary to solve the mathematical problems.

Group and class discussions were widely used in the sessions. Students were encouraged to exchange ideas with each other and construct their own knowledge with facilitation by the teacher. For each session, the teacher would introduce and explain the key processes involved in problem solving. The teacher also assumed the role of facilitator which required him or her to scaffold and guide students' learning. Students were also given opportunities to verbalise their learning processes by sharing the reasoning behind their attempts to solve the problem.

At the end of the MProSE lessons, the students were given a Practical Test. They were given one hour to individually answer a problem on the Practical Worksheet. The problem that we used for this test is shown in Figure 1.

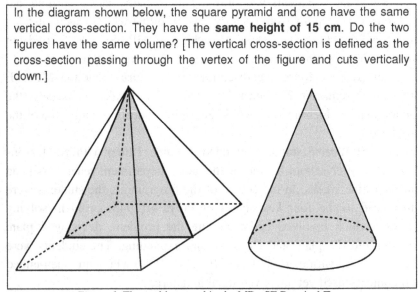

In the diagram shown below, the square pyramid and cone have the same vertical cross-section. They have the **same height of 15 cm**. Do the two figures have the same volume? [The vertical cross-section is defined as the cross-section passing through the vertex of the figure and cuts vertically down.]

Figure 1. The problem used in the MProSE Practical Test

3 Data Collection and Analysis

Throughout the first implementation, data were collected to guide the school in making improvements for its second implementation and to assess the impact of the initial teaching of MProSE on learners. In particular, we collected students' test scores in the Practical Test and students' feedback about the programme through an online survey. We also obtained a more in-depth account of the experiences of two students who underwent the course. Teachers also shared their reflections and weekly discussions with the NIE researchers were conducted to form part of the process of evaluating the programme.

Students' MProSE Practical Test Scores. Table 1 provides a summary of the students' test results in the MProSE Practical Test. We found that the average achievement of students for the MProSE test was 81% . Furthermore, we also noted that one in ten students were able to attain full marks in their MProSE Practical Test. In a class-by-class analysis, there was even one class where three in ten students achieved full marks in the MProSE test. We found these results encouraging since the MProSE rubric placed premium on the students' thinking processes and their efforts to extend their thinking through going through Pólya's fourth stage. Thus, their performance in MProSE reflected how students were beginning to develop more holistically by moving away from exam-oriented learning.

Table 1
Summary of Students' Test Results

Average percentage of students' MProSE Practical Test scores	81%
Percentage of students who obtained full marks in the MProSE Practical Test with full scores	11.3%

We also wanted to look into how the "remedial" students performed. In TKGS, students who did not pass their final year exam in the previous years are placed in the remedial programme to better support their learning at the next grade. One of the key concerns we had when implementing MProSE was its effect on students who struggle with mathematics based on their test results in past examinations. It remained to be seen if these students would be able to handle MProSE. Based on the MProSE Practical Test, the remedial students' average score was

76.3% while the average of the rest of the students was 82.6%. We have no baseline data to compare if the gap between the remedial group and the rest had widened or narrowed. But we judge 76.3% to be a good performance for the former group and perhaps an indicator of the premise that MProSE can indeed be for *everyone*.

Students' Survey Results. Besides looking at whether students of different abilities can follow the MProSE programme, the school also conducted an online survey to obtain feedback from students about how they felt about the MProSE module. We used the survey questionnaire found in Appendix D which was provided by the NIE researchers.

Students were asked to complete the survey after school by their respective mathematics teachers. Of the 280 students in the cohort, 201 students (or about 72% of the students) completed the online survey, which was conducted at the end of the MProSE programme, but before the administration of the MProSE Practical Test. When we analysed the survey results, we paid particular attention to five items which asked the students to respond to the following statements using a 6-point Likert scale (with 1 corresponding to Totally Disagree and 6 corresponding to Totally Agree):

- Posing related problems (by extending, adapting or generalising) is a part of learning mathematics.
- Pólya's model of problem solving is useful to me.
- The Practical Worksheet guides me in applying Pólya's model of problem solving.
- I did not enjoy solving the problems presented in this module.
- The way of thinking taught in this module will help me in other mathematics modules.

We focused on these five items because we found they pointed directly to shifts in the students' mathematical thinking and affect towards mathematical problem solving. Table 2 provides the percentage of students who responded with agreement (i.e., gave a rating of 4 or higher on the Likert scale) for each of the statements.

The overall results showed that a majority of students gave

favourable responses to four out of the five statements. Such results indicated how the programme had a positive effect on the students. The most encouraging piece of data was how 74% of the students agreed that the Practical Worksheet guided them in applying Pólya's model. On the whole, while the survey results point towards some degree of success, we could improve helping *more* students appreciate the problem solving processes that the MProSE programme desired to inculcate in students.

Table 2
Summary of students' responses to certain survey items

Statement	% of students agreeing with the statement
Posing related problems (by extending, adapting or generalising) is a part of learning mathematics.	58%
Pólya's model of problem solving is useful to me.	55%
The Practical Worksheet guides me in applying Pólya's model of problem solving.	74%
I did not enjoy solving the problems presented in this module.[a]	57%
The way of thinking taught in this module will help me in other mathematics modules.	59%

[a] A favourable result for this item requires students to disagree with the statement.

The open questions in the administered survey provided us with a better sense of what we were able to make the students value or fail to value. It was encouraging to note from students' comments and feedback that they had unknowingly shifted their focus from routine to process. In particular, instead of focusing on application of mathematical formulae, many students recognised the value of understanding the problem and use of heuristics for one to be able to solve mathematics problems successfully. For some, Pólya's model was valued "as it teaches in some way not to give up on a question" as one student responded. Students also recognised that possessing virtues such as patience and perseverance were also essential in mathematical problem solving.

Students found the process of solving the problem as one of the best parts of the module. This was particularly true when they managed to solve problems they had never seen before. They also found satisfaction in applying different heuristics to explore a problem from different perspectives. Tackling interesting questions and working in groups also played a big part in the success of the module. For example, one student stated that "the interaction with [her] peers as [they] learnt from each other while brainstorming and working on the various methods of getting to the answer" was the best part of MProSE.

Students' responses to the question, "Which was the worst part of the module?" gave us a glimpse of where we fell short in terms of imparting the essence of problem solving to our students. A word cloud of the students' responses is provided in Figure 2. In this word cloud, the font size of a word is made proportional to the frequency of the word's occurrence in a text. In particular, the larger the font, the more frequent the word appeared. While it does not provide a precise measure, it highlighted for us the various areas that we should find time to address in the future.

Figure 2. Word cloud of students' responses to the question, "What was the worst part of the module?"

Understandably, Pólya's fourth stage of checking and expanding was perceived as the most difficult part of the module. In addition to this, a lot of students found following Pólya's stages tedious and boring. On top of all this, students found the timing and setting of the sessions not conducive for learning. As one student puts it, "We had to stay back after school for the module when the weather is so hot and we were all tired so we couldn't concentrate."

Student Accounts. During an informal feedback session conducted at the end of the MProSE programme, teachers found that students who showed positive responses were largely those who were able to experience the "eureka moment" – the thrill of solving a problem which they had earlier deemed to be beyond their level of comfort. Indeed, these students who experienced this had a boost in their level of confidence and interest in mathematical problem solving. As such, they showed improvement in their attitude during mathematics lessons and were more receptive towards challenging questions. In order to better understand and appreciate the experience of these students, we interviewed a couple of students who we felt represented this set of students well. The first student, Jane,[1] described her MProSE journey in her after-thoughts as follows:

> I felt that in the beginning of the MProSE training course, it was slightly dull and uninteresting because many other students made negative comments – nobody wanted to stay back for extra lessons especially after a long day of school. But on the fourth, fifth week of the training, the lesson on "Lockers Problem" caught my attention as it involved a practical situation instead of the common problems we face in exams that can only be solved on paper. Subsequently, I paid more attention in class and gained more interest in mathematics, which changed my perspective towards problem-solving. Problem-solving is no longer based on a rigid formula. Instead, I am encouraged to think of alternative

---

[1] Student names appearing in this paper have been changed to protect their identity.

solutions – this has engaged me in learning mathematics. In fact, the best part of MProSE was getting the hang of solving the problem after understanding what MProSE is all about!

It was earlier pointed out that Pólya's fourth stage of checking and expanding was found to be one of the most difficult for students to appreciate. Ironically, students who experienced the eureka moment spoke positively of this component in the Practical Worksheet wherein they were guided to reflect on their own thought processes whilst coming up with alternative solutions. In particular, they derived a greater sense of satisfaction when they could come up with alternative solutions to the problems. The systematic approach towards problem solving alongside the emphasis on flexible thinking allowed some students to re-establish their confidence towards problem solving and these students were able to realise that solving mathematical problems can actually be fun and interesting. One such student was Sally who used to give up quite easily when faced with tough mathematics sums. She reflected that her perspectives towards problem solving changed after going through the MProSE programme, and she was certain that her mathematical problem solving skills and processes will stand her in good stead for the future.

MProSE has inspired me to always think of new ways to solve a problem and I think this flexible thinking would help me in my examinations. I feel very satisfied when I get to solve a problem and then feel very confident when I still get the correct answer after checking a few times.

Perhaps it is significant to mention that Jane and Sally did not perform in the standard mathematics examinations as well as their peers. Table 3 provides a summary of some of their examination scores compared to their MProSE Practical Test score. Despite the fact that their examination scores fell below the cohort's mean, both Jane and Sally still managed to get full marks in the MProSE test. These are not exceptional cases. The students who had done well for their MProSE examination were not necessarily those who had aced their mathematics examinations. There are too many factors that could have contributed to the results mentioned, such as students' motivation, teacher's encouragement, and home support. Interviews with these students revealed the experience of "eureka moments" during their MProSE journey and strong support and encouragement from their teachers as contributing to their positive affect. Notably, these students were persevering in their learning attitude and did not give up easily. This was yet another good indicator that mathematical problem solving can be manageable for all students regardless of their mathematical ability based on their test scores.

Table 3
Jane and Sally's performance in standard mathematics examinations and in the MProSE Practical Test compared to the cohort's mean

	Jane's marks	Sally's marks	Cohort's mean
PSLE[2] Mathematics Grade	B	B	----
Percentage score in the Year 7 final term examination	58.3	41.3	64.3
Percentage score in the Year 8 mid-year examination	54.0	36.0	56.6
Percentage score in the MProSE Practical Test	100	100	81

Teachers' reflections and discussions. While there were pockets of positive outcomes in our first implementation of MProSE, overall, we felt that we could do much more to improve the way MProSE was being carried out in the school. We recognised that among the factors that contributed to this mediocre implementation were the following:

---

[2] PSLE stands for Primary School Leaving Examination. It is taken by all Singapore students at the end of their Primary School (Year 6).

- *Schedule and classroom setting.* MProSE was conducted after school hours which the students did not welcome.
- *Teacher preparation.* Teachers were actually given considerable support to help them carry out the MProSE lessons. Support included the provision of the initial training with the NIE researchers prior to the implementation, and the setting aside of common time for the teachers to work through the problem and the lesson during the implementation. However, owing perhaps to the newness of the programme, we recognise that teachers may have required more support and preparation; more time and space should be given to internalise the teaching of MProSE. Teachers also needed time to build rapport with the students as some of them do not teach the level.
- *Competing use of resources after school.* Because MProSE was conducted after school, some students had to straddle between the MProSE class and other important programmes such as external competitions. Some teachers sometimes also found themselves needing to be involved in other programmes.

We believe that these factors contributed to the weak buy-in among some of the students. However, despite the challenges faced and weaker-than-expected outcome after the first implementation, the teachers themselves learnt from the entire experience. And teachers saw value in persisting in helping our girls enjoy the problem solving process and appreciate the power and beauty of mathematics through MProSE.

The nature of the task and student participation matters. The nature of the tasks involved was found to be crucial in helping students develop better problem solving skills. The mathematical tasks had to be crafted in a manner such that students were encouraged to take on a flexible approach towards problem solving. On the other hand, teachers had to ensure that the level of difficulty was within the students' comfort levels and that opportunities for students to explore alternative solutions were not impeded. Teachers also noticed that when tasks incorporate real-world contexts, students felt more interested in solving the problem as they were able to relate the problem to their own experience.

In addition, teachers also noticed that students who were able to contribute their ideas and explanations during their group and classroom discussions derived a greater sense of satisfaction and meaning than those who did not participate as actively. Hence, these students had more confidence in themselves towards solving mathematical problems.

The Practical Worksheet can promote deep understanding. It was heartening to note that teachers who conducted the MProSE lessons agreed that the Practical Worksheet was a tool that can promote deep understanding and mathematical learning. Learners had to practice communicating their understanding, and make their thinking visible. By going through Pólya's stages and making the necessary loops, students were trained to consistently look for alternative solutions and to strive for accuracy and precision. Adapting, generalising, and extending a problem were perceived as creative processes that developed flexible thinking as students learnt to paint what-if scenarios.

The Practical Worksheet served as the tool that tracked the skill and affective development of learners. The ease of use of the guiding questions acted like a compass that directed the learners through the problem solving journey till a 'habitual path' was formed mentally. This seemingly forced adoption of a systematic approach towards problem solving especially benefited the students who often found themselves unable to decide how to proceed in mathematics. It gave the learners handles on what to do next and helped the learners take productive steps forward.

MProSE can develop character and build lifelong skills and appreciation for mathematics. Besides skill development, MProSE took care to build self-esteem and confidence through its assessment rubrics design. The use of problems that stretched beyond the current syllabus meant that the solutions were usually not apparent. The emphasis on process in awarding of marks for assessment gave many chances for students to fail and try again until a correct solution was reached. This in turn trained persistence and raised interest in mathematics as learners were not being penalized for failed attempts. Creative solutions to the problem could begin to unfold as a result. The assessment criteria implicitly taught delayed gratification, deferred judgment and thoughtful reflection towards solving problems as they were forced to pen down their plans and strategies before actually formalising their solutions and conclusions.

Teachers also found students who recognised that competence in mathematics did not only mean getting the correct solution, but rather, it also meant understanding and grasping the mathematical processes and strategies involved as they attempted to solve these problems.

Though we cannot claim that our first implementation was a total success, it made us grow more confident about how MProSE can truly equip our students with essential skills for the 21st century. We thus took to preparing for the second implementation of MProSE in TKGS.

4 Second Implementation – The Journey Ahead

The data we collected in the first implementation gave ample insights into what we have done well and what we could improve on to make MProSE work better for our students. For the second implementation, we particularly wanted students to positively view those aspects of the programme which were perceived negatively in the first implementation. Thus, we made the following accommodations and improvements:

- Instead of holding the MProSE classes after school hours, an additional 50-minute teaching period was planned into the school timetable to teach MProSE in the second and third term[3] for 10 weeks.
- All MProSE classes are encouraged to be held in a large, fully air-conditioned room so that students will have enough space to discuss and engage in problem solving. We also encourage group seating arrangement.
- Manipulatives or props appropriate for each problem were created and were made available in the MProSE room for teachers to use, and for students to be able to more meaningfully

---

[3] TKGS's school calendar is divided into four terms.

understand and explore the problems at hand.

- A booklet for students' use was created and designed to capture their critical processes and eureka moments. This is intended to help students better recognise the connections in their learning and better recognise the value of problem solving. Selection of the problems took into consideration their difficulty level, their appropriateness for the students' cohort, and their level of tie-in with the year level's mathematics curriculum syllabus.

- Detailed lesson plans were drafted and fine-tuned for teachers to use. The NIE researchers also provided their inputs.

- Facilitation cards were adopted to help teachers manage the large number of groups in each class. These cards, suggested by the NIE researchers, serve to help teachers provide students with the appropriate scaffolding response according to the level of progress that the students demonstrate.

With each implementation, we overcome some challenges but at the same time, we anticipate new challenges ahead. But through the commitment of the teachers and the NIE researchers, we trust that we will persevere.

5 Conclusion

MProSE equips students with essential problem solving tools and processes. This is central to a student's confidence towards mathematical problem solving. Students' involvement in MProSE tasks allows them to apply their acquired mathematical skills which are essential to a child's development of critical and flexible thinking and cultivation of life-long habits of mind.

MProSE, in essence, is a brave attempt at helping teachers teach how to solve problems. It is a module that powerfully equips learners to learn to create, collaborate, and think critically in order to be ready for an unpredictable future. It is transformational for teachers and students – from superficial, occasional use of Pólya's stages to an in-depth application of Pólya's stages for solving problems. In the words of one of the teachers, "MProSE can level up students' problem solving abilities as it provides a structured programme for students to return to the fundamentals of mathematical problem solving by introducing Pólya's four-step problem solving framework and using problems that are a little beyond the students' current curriculum syllabus to learn beyond the life of tests to tests of life."

Comments from the NIE Researcher.

The objective of Mathematical Problem Solving for Everyone (MProSE) resonated early with the objective of the mathematics department of TKGS to "nurture creative problem solvers and self-directed learners". From this shared beginning, we have worked with a group of teachers who are enthusiastic and willing to learn to help their students discover and realise 'real' mathematical problem solving. Without these attributes, it would be impossible to implement the necessary features of our problem solving package – that it be within the main curriculum, that its assessment be valued, and that all 4 stages of Pólya's model be learnt. The article above states that the teachers have also grown as a result of the project, a fact that we can confirm after conducting an initial problem solving workshop for them and after working closely with them through a year of lesson discussions and module design/redesign.

We are confident that MProSE will be successful in TKGS given the strength and quality of the mathematics department and its leadership. There are, however, two areas which I feel that we must tread carefully.

The first is to not be overly optimistic about what MProSE can achieve. Although 57% of the students agreed that they did not enjoy the module (which, in the light of the other generally positive responses, could be positively explained as that they interpreted it to mean that they found it 'challenging'), the problems and thrust of the module should not be changed too drastically to create an easy and non-threatening atmosphere. Problem solving by its nature is challenging and oftentimes stressful. The key is to regularly remind the learners that the module is to teach them what to do when they are 'stuck' and not merely to make all the problems palatable. In addition, it would be wise to consider Schoenfeld's (2011) view that a teacher's goals determine her actions in the classroom. Unfortunately, if there are too many goals to achieve within the short amount of time afforded in each lesson, the teacher may even fail to achieve any. More likely, however, is that the teacher will prioritise her goals. But since this will most likely interact with her deeply held beliefs, problem solving goals new to her may be inadvertently set aside as a result.

The second is that teachers should leverage on their enthusiasm for their students to quickly master for themselves the problem solving process. We have found that those who are most enthusiastic about propagating Pólya's model are those who have successfully used it to solve their own mathematics problems. To this end, I would strongly encourage the teachers to try to solve mathematics problems that they encounter or that have been posed to them. An online Professional Learning Community (PLC) for mathematical problem solving has been set up for all the schools in MProSE. Interesting problems are posed in the PLC and solutions are written in an adapted form of the Practical Worksheet. We believe that there will be a quantum leap in the teaching of MProSE lessons when the teachers themselves master problem solving.

Acknowledgements

The authors would like to thank

Subject Head Diane Ang Koh Mei Lin, together with 7 committed teachers as well as the rest of the mathematics department team who took on the challenge to tread the unknown path to improve students' problem solving and showed students the way to learn;

the 2012 TKGS Year 8 students who took part in the survey and gave us insights to improve mathematics education; and

School leaders Ex-Principal Phyllis Chua-Lim, Principal Mary Wong-Seah, Ex-VP Marilyn Chia, Ex-VP Tan Ming Fern, VP Lee Beng Choo, VP Shirley Wong Kah Wai, VP New Yi Cheen, and the school management committee for their support throughout the whole implementation process.

Chapter 5

Mathematical Problem Solving in Jurong Secondary School

TEO Oi Mei   MOHAMED Faydzully On Bin Mohamed Razif
OOI Wei Yong   Simranjit KAUR   TAN Siew Hong
Jaguthsing DINDYAL (NIE Researcher)

In this chapter, we document the implementation of a problem solving module in the Year 7 Express classes in Jurong Secondary School. We highlight some of the factors that supported the implementation of the module as well as some of the difficulties that we faced. Lesson 9 on the problem "Intersection of Two Squares" is used as an example to illustrate a typical lesson.

1 Motivation to Participate in MProSE

Due to the overall good performance of Singaporean students in the Third International Mathematics and Science Study (TIMSS) in 1995, the Singapore mathematics educational system has been getting a lot of attention in the international education community (Lim, 2002). We think the success in mathematics achievement was mainly due to teachers' emphasis on the skills and concepts components of the pentagon of the Singapore Mathematics Curriculum Framework (see Figure 1 in Chapter 1 of this volume, p. 2).

An over-emphasis on skills and 'drill and practice' however can be dangerous as it can undermine students' problem solving abilities when they are faced with authentic problems. Students may become proficient only in solving routine problems but may be at a loss when faced with more unconventional problems. Therefore, it is also essential for the teachers to plan learning environments where the students develop multiple problem solving strategies rather than use a single known approach when solving mathematical problems. In other words, teachers should consciously teach students the processes of problem solving. With these in mind, we at Jurong Secondary School decided to embark on a journey to help our students become more proficient in using mathematical problem solving skills in their daily mathematics lessons. This journey was made possible through our participation in the Mathematical Problem Solving for Everyone (MProSE) Programme.

2 Planning and Preparations for the Problem Solving Module

The Mathematics Department of Jurong Secondary School worked with researchers from the National Institute of Education (NIE) to implement a research programme called MProSE in the first semester of 2012. The purpose of the programme was to better equip the Year 7 Express[1] students with mathematical problem solving strategies which they could apply in their learning of mathematics in the classroom and beyond, thus, making them more flexible in using their thought processes. This was based on a shared vision that after equipping students with problem solving strategies based on a sound problem solving model (in this case, Pólya's (1945) model), students would be able to utilise the problem solving processes to acquire new mathematical knowledge. The aim was to augment the standard drill-and-practice for procedural fluency. Also, it was expected that the collaboration with the researchers would benefit the teachers professionally as they would improve their teaching of mathematical problem solving.

The Principal was supportive of the mathematical problem solving project and expressed his belief that the approach of teaching problem solving as proposed by the NIE researchers could be the direction for the school's mathematics department. This commitment was evident in the

---

[1] In Singapore, secondary students are channeled into three ability streams – Express, Normal (Academic), and Normal (Technical).

allocation of ten hours of curriculum time for the MProSE lessons for all the Year 7 Express stream students in the school. Furthermore, the grades obtained by students on the test conducted at the end of the problem solving module were to be incorporated as a component of their continuous assessment in school.

The school decided to make use of the set of problems as proposed by the NIE researchers instead of adapting or modifying the problems. Teachers, being new to teaching problem solving skills in class, had to invest time to understand how to teach the skills using the problems suggested by the NIE researchers. In adopting the problems as proposed, the teachers were able to fully concentrate on teaching the four stages of Pólya's model instead of having their focus diverted towards the adapting problems. The plan was also for the teachers to make some of their recommendations to the NIE researchers on the appropriateness of the problems with respect to the school's context at the end of the first implementation of the programme.

There were two main issues that needed to be sorted out before the four mathematics teachers who were earmarked for teaching the module could conduct the lessons. The first challenge was the need to increase the curriculum time. This was to ensure that the completion of the school mathematics syllabus would not be affected because of the teachers' involvement in conducting the problem solving MProSE lessons. With strong support from the school management, one additional hour was added to the timetable in the first semester.

The second issue was the teachers' readiness to conduct the rather 'unconventional' problem solving lessons. Teachers attended a three-day problem solving professional development workshop in November 2011. However, they were still unsure of how to conduct the lessons themselves. To allow the teachers to have more time for in-depth discussion of the lessons together, one hour per week of Professional Development (PD) time was allocated. The NIE researchers, who were also the consultants for the problem solving lessons, were occasionally invited to attend the PD time to help to clarify certain doubts on the pedagogical approaches as stated in the lesson plans. This helped increase the teachers' confidence in teaching mathematical problem solving skills and ease the teachers' fears of conducting the lessons. A total of seven sessions were used to prepare the teachers and the first MProSE lesson was conducted in eighth week of Term 1.

It was also decided that the Head of the Mathematics Department (HOD)—the first author, hereafter referred to in the first person—would take the lead in teaching the problem solving lessons. That is, I taught the problem solving lessons of one class even though I was not a resident teacher of any of the Year 7 Express stream classes. Furthermore, the schedule of the MProSE lessons was adjusted so that the Year 7 teachers could observe my lesson before they taught their respective classes. I took on this task being cognizant that the teachers might not yet be comfortable with teaching problem solving. I had also always grappled with how best to teach mathematical problem solving according to Pólya's model; this research provided me with the insight into how this could be done and the opportunity to carry it out.

3 Implementation of the Problem Solving Module: Perspective of the HOD (Mathematics)

The explicit teaching of the problem solving module to the Year 7 students started in January 2012. The whole package of lessons consisted of ten one-hour lessons. A detailed description of the lessons can be found in the MProSE guidebook for mainstream schools. These lessons, which were incorporated into the current mathematics curriculum, were held once a week.

During the lessons, students were introduced to Pólya's problem solving model consisting of four stages: understand the problem, devise a plan, carry out the plan, and check and expand. Pólya's model was often reiterated in the lessons and students were highly encouraged to follow the four stages whenever they encountered any mathematical problem. Different heuristics were also introduced to aid the students whenever they got 'stuck' in a problem. The teacher would commend students who tried to use heuristics to solve mathematical problems in order to encourage them. Most importantly, it was emphasised that students articulate their thought processes in words to help inculcate desirable mathematical problem solving habits in students.

Lessons usually started with the review of the previous lesson and a short five-minute discussion on the steps in Pólya's problem solving model. Subsequently, the students were asked to solve a Problem of the Day (PoD) and were given 30 minutes to complete it. Although every student was expected to hand in their own individual work, they were allowed to discuss in pairs. The role of the teacher during this section was to facilitate the understanding of the problem and ask leading questions, being very careful not to give away any mathematical step or answer to the students. After about 30 minutes, the teacher would proceed to discuss the PoD by selecting two or three students to share their answers. The teacher made it a point to emphasise the proper steps and thought processes rather than mere answers. In the last five minutes, the teacher would conclude the lesson and the homework for the day would be assigned.

The NIE researchers observed Lessons 1 to 4 of my classes wherein I adhered closely to the proposed lesson plans and problems. The students' responses to the problems were more encouraging than expected with many students attempting to use algebraic notations to represent mathematical entities, and some students attempting to use an algebraic equation to solve the problem (even though they had not learnt solving algebraic equations at this juncture).

With regard to Pólya's stage 4, I expressed my concern that Year 7 students might have difficulty with too many sophisticated extensions, generalisations, or adaptations as proposed in the "Check and Expand" section. Thus, in Lesson 4, I engaged students more on the checking of the reasonableness of their solutions rather than focus on expansion.

4 Zooming-in to a Sample Lesson

In Lesson 9, I modelled and solved the homework problem "Intersection of Two Squares" on the board. The problem is stated below.

Two squares, each s on a side, are placed such that the corner of one square lies on the centre of the other. Describe, in terms of s, the range of possible areas representing the intersections of the two squares.

The board was divided into four columns, labelled Stage 1, 2, 3, and 4. As the objective of this lesson was to highlight the control decisions made by a problem solver, I 'thought aloud' by verbalising 'my thought processes'[2] in the classroom as we tackled the problem. This was to demonstrate that a good problem solver had control in her decision making. Furthermore, the model lesson was conducted for students to reiterate how they were to fill in the worksheet for the four stages.

In Stage 1, I asked the students which parts of the problem they had read and not understood. I underlined the key words of the first sentence as follows: "<u>Two squares</u>, each s on a side, <u>are placed</u> such that the <u>corner of one square lies on the centre of the other</u>." I then repeatedly asked the students what they understood by this sentence. After that, students were led to apply the heuristics, *draw a diagram* and *consider a simpler problem*. I specifically asked the students how I should draw the two squares in order for it to become a 'simpler problem'. I then started drawing a few versions on the board, leading the students to identify the simplest diagram that we could start with.

Once the students were able to visualise the problem themselves, I verbalised my thought processes loudly by asking myself some *control* questions:

1. What (exactly) am I doing? Clearly describe what I am doing NOW.
2. Why am I doing it? Clearly describe what I am doing in the context of the BIG picture – the solution. Clearly describe what I am going to do NEXT.

In the second column of the board under Stage 2, I shared with the students that I was clear with what I wanted to do and this was to rotate one square and fix the position of the other. This became my plan. I drew the diagram for the general case under Stage 1. I then asked the students to find the area of the first simple case (i.e., $\frac{1}{4} s^2$). The corresponding

---

[2] Here, 'my thought processes' do not refer to the actual thought processes that I employed in problem solving; rather, I was modeling for students according to Pólya's stages.

board work for Stages 1 and 2 are shown in Figures 1a and 1b respectively.

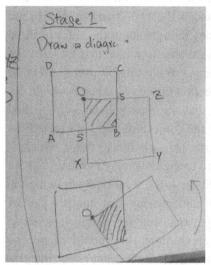

(a) Stage 1: Understand the problem

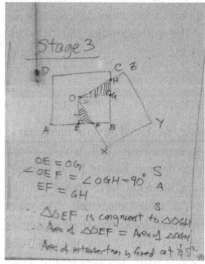

(c) Stage 3: Carry out the plan

(b) Stage 2: Devise a plan

(d) Stage 4: Check and Expand

Figure 1. Photos of how the four stages of the problem in the sample lesson were presented on board

After clearly articulating my plan, I moved on to the third column for Stage 3: Carry Out the Plan. I drew the diagram representing the general case. I then labelled the vertices and the intersection points of the two squares as shown in Figure 1c. I shaded the two triangles $\triangle OEF$ and $\triangle OGH$ and asked the students what they thought about the areas of these two triangles. Most of the students were able to notice that the areas of these two triangles 'looked' the same. With this, I asked them to conjecture what the area of the intersection of the two squares would be. Some students immediately shouted that the area will be $\frac{1}{4}$ s^2. I then asked the students to try to show that the conjecture will hold regardless of the position of the squares relative to each other. After which, I showed them how to prove that the two triangles were congruent.

On their own, students checked (Stage 4) the solution by drawing the 'rotating' square in another quadrant and proving that another pair of triangles was congruent. In applying expansion for Stage 4, students were encouraged to think further about how they could adapt the problem. Some students gave ideas of changing the squares to triangles or other polygons while I shared that they could consider two regular hexagons as an example of an adaptation. For further extension, I also suggested that students consider a 3-dimensional cube instead of a 2-dimensional square. Figure 1d shows the corresponding board work provided for Stage 4.

5 Reflections

During the implementation of the module, we faced several challenges. Firstly, near the end of Lesson 8, students were still having some difficulties with the Practical Worksheet despite having been through eight lessons on how to use them. Some students were found writing their solutions in Stage 1 or Stage 2 for their homework. This was probably due to the fact that they were not used to writing their strategies explicitly given that their usual practice was writing only the solutions for mathematics problems. To overcome this issue, teachers had to write down the solution explicitly as illustrated in the sample lesson above to demonstrate how to fill in the Practical Worksheet. This was carried out so that students would have a clear model on how to use the Practical Worksheet.

The ten MProSE lessons ended towards the end of April which coincided with the revision period for Mid-Year Examinations (MYE). Due to this schedule constraint, the teachers were not able to conduct the MProSE Practical Test before the MYE. Thus, the test had to be conducted after the MYE. As such, we felt the need to conduct an extra MProSE lesson prior to the Practical Test. This extra lesson was essentially a revision to refresh students' memories about how to apply Pólya's model and use the Practical Worksheet. Furthermore, the extra lesson also familiarised the students with the MProSE assessment rubric (see Appendix C) for the Practical Test. To carry out the latter, a sample filled-up Practical Worksheet was made which demonstrated how a student could go back to Stage 2 if a plan failed in Stage 3. We went through the assessment rubric using this sample as a guide. Through this exercise, we wanted to assure students that the marking did not simply consider final answers, but also the problem solving process that they undertook. Thus, even if they arrived at an incorrect answer, they could still get a passing mark as long as they satisfactorily carried out Pólya's model and showed this on their worksheets.

Another challenge we faced was that we were not familiar with marking using the given rubric. Because of this, the NIE researchers were consulted a few times to make sure the Practical Tests were properly marked. After the consultation sessions, we were then able to calibrate the marking among the teachers and we subsequently gained confidence in marking using the rubric.

During the overall implementation of the problem solving module, teachers felt highly dependent on the resources provided by NIE and often needed more time to prepare for and discuss the lessons before class. In class, teachers often felt unsure of how much information to provide students when they asked questions. The limited contact time with the students for problem solving was also an issue. The teachers felt that the one-hour slot for reviewing homework, discussions and trying out the PoD was not enough for them to provide quality feedback to the students. Many felt that it might have been better if students could attempt fewer questions and the teachers spend more time discussing the problems and students' thought processes qualitatively.

6 Preparations for the Next Round

Some of the lessons in the package were found to be not suitable for our students. For example, the problem "Going for a Movie", as reproduced below, was found to be too easy as the students had many similar examples on the topic for highest common factor.

> *Going for a Movie Problem.* On a Tuesday, Alice, Bernice and Carol and Dory, met for a movie. After the movie, they made plans for the next gathering. Alice, Bernice and Carol said that they could only go to the movies every 6, 3 and 4 days respectively, starting from that Tuesday. Dory said that she could go to the movies every day except on Fridays. After how many days would the four friends be able to meet again for a movie?

On the other hand, the problem "Let's Draw a Graph" below was too difficult for the students as they had not learnt how to sketch a step-function graph before.

> *Let's Draw a Graph Problem.* In a car park, it was advertised as such:
>
> > First hour parking: Free
> > Subsequent parking: $1 per hour or part thereof
>
> Sketch a graph for parking fee versus the number of hours of parking with axes as shown below.

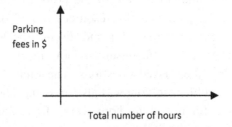

Thus, the NIE researchers will be working with us to fine-tune the problems so that they will better match our students' level and needs.

As part of succession planning, our school will deploy two or three teachers who were not part of this year's implementation to teach the MProSE lessons to the new batch of Year 7 students in the next implementation. The plan is for two teachers to pair up and co-teach ten lessons. Each pair will consist of one MProSE-experienced teacher and another who will be doing it for the first time.

The school will also infuse the problem solving Practical Worksheets in the mathematics lessons of the Year 8 Express stream students. This will serve as a continuation of this year's efforts to develop the students' problem solving skills. The Year 8 lessons will consist of a PoD using the Practical Worksheet, which will be related to one of the topics taught in the Year 8 mathematics curriculum. The problems will be provided by the NIE researchers. Students will then be tracked on how well they perform in their school examinations after going through the MProSE modules for two years.

7 Conclusion

With the experience gathered this year, we have gained much confidence in teaching the module. This was especially so when we received positive feedback from students. Most students gave feedback that they enjoyed the lessons as they stimulated their thinking. The skills they learnt helped them with other mathematical problems. Although a small fraction of the students were still fearful of the unconventional problems, these students were generally receptive towards learning the skills for mathematical problem solving. This year has been a fruitful journey for the teachers and we are ready to embark on the second year of our journey in MProSE with full support from the school management and the NIE researchers.

Chapter 6

Mathematical Problem Solving
in Bedok South Secondary School

Kavitha THAVEN   LOH Jia Perng   Samuel GOH De Hao

LIM Ming Ming   LIM Yih   Kavitha BUTHMANABAN

QUEK Khiok Seng (NIE Researcher)

In this chapter, we describe our implementation of the curricular innovation called Mathematical Problem Solving for Everyone (MProSE) at Bedok South Secondary School. We also share our collective reflection of the process we undertook. In the first part we share our motivation for taking part in the programme. We then describe how it was implemented in the school. Finally, we reflect on our involvement in the programme, highlighting aspects that came to our attention, describing its strengths and the areas that we might want to execute differently for the next programme trial.

1 Motivations and Aims

Bedok South Secondary School came into the Mathematical Problem Solving for Everyone (MProSE) programme with the aim of improving students' learning experience in mathematics as well as enhancing teachers' capacity in teaching mathematical problem solving. While the students' academic results in mathematics have been improving, they struggle to retain the concepts that they are supposed to have learnt. Furthermore, they demonstrate a lack of depth in their understanding of these concepts.

Students seem to see mathematical ideas as disconnected and

unrelated to one another. For example, algebra is often treated as a collection of rules and procedures with little link to numbers and arithmetic. Topics such as mensuration, profit and loss, rate, ratio and proportion seem to stand alone individually as a series of formulae to be applied in each respective context instead of being viewed as an extension of proportional reasoning. While a minority of students may be able to see the links between some of these concepts, the majority lacks the habit of mind and the motivation to seek the underlying links. Hence, MProSE's emphasis on the mathematical problem solving process aligns itself well with meeting the students' needs for

- knowledge of the mathematical problem solving process, making it possible for them to have greater metacognitive awareness of their own thought processes during mathematical problem solving;
- appreciation of mathematical problem solving and building their sense of self-efficacy in solving mathematical problems by engaging in successful problem solving; and
- developing the habit of mind and the motivation to seek links among the mathematical ideas that they come across.

Underperforming students usually lack the skills and motivation to seek links among mathematical ideas and to engage effectively in mathematical problem solving. As their teachers, we (the authors of this chapter) saw the need to help them develop their capacity and motivation in order to help them improve their academic performance.

To accomplish this, we recognised that there was a need for us to enhance our own capacity to engage in mathematical problem solving so that we could better teach it to our students. The mathematics teachers in our school came from diverse academic backgrounds. A few of us majored in mathematics, but most of us received our degree in other fields such as engineering, accountancy or computer science. Thus, our experiences in applying mathematical concepts varied and we may have different understandings of what mathematics is and its purpose. Our common denominator was our successful experiences in working with the mathematics that is tested at the standardised examinations. However, just like many academically successful students, we may not be fully aware of our own thought processes to be able to apply them to unfamiliar situations let alone delineate our thought processes explicitly enough to explain and coach others into achieving the same.

In addition, we wanted to enhance our capacity to facilitate students' learning to engage in mathematical problem solving. The scaffolding might take the form of teacher questioning, timely and appropriate teacher intervention, direct instruction, giving students problems appropriate to their ability, grouping students by ability and pacing students' learning experience. Hence, we also participated in the programme to

- enhance our own experiences in mathematical problem solving, and to raise our awareness of the processes, and
- increase our pedagogical repertoire to better facilitate these learning experiences for the students.

Comments from the NIE Researcher.

Educational research has been criticised for its lack of efficacy at school level. Research users, here teachers and schools, are right in demanding sufficient ecological validity in the research recommendations to assure the potential of success in their implementation. As we see in the preceding recollection, ecological validity by itself is not enough to translate research into practice. A potential research user must see a 'relative advantage', to borrow a term from Everett Roger's Diffusion of innovations (2003). Bedok South Secondary School sees the MProSE design for mathematical problem solving as a means to meet their needs. A confluence of several factors, e.g., student needs, teacher professionalism, school leadership support, had to be in place before a curricular innovation like the MProSE design could find a foothold in schools.

2 The Implementation

Choice of students. In Bedok South Secondary School, the MProSE design was implemented for 120 Year 7 Express[1] students and 10 students from the Year 7 Normal (Academic) stream as the latter group was also using the syllabus for the Express stream for their mathematics lessons between Term 1 and Term 2 of 2012. Year 7 students were selected as we believe that, being new to the school, they would be more receptive to changes from the usual curriculum. Starting with Year 7 students also allowed us to have more opportunities to work with the students over their secondary school years, as we believe that these skills and beliefs need to be developed over time.

Teachers' preparation. The preparation for teachers to carry out the programme was done in two stages. Before the launch of the programme, we attended a 10-hour workshop conducted by the researchers from the National Institute of Education (NIE). We were given problems similar to the ones that the students would be using, albeit set at a slightly higher level of difficulty so that we had the opportunity to experience the struggle that students would have to go through. The workshop also allowed us to experience the kinds of scaffolding that might take place during the lessons.

During the implementation of the programme over the school year, we provided mutual support for each other through the conduct of weekly one-hour team meetings exclusively meant for teachers to prepare for teaching the MProSE lessons. The team consisted of Year 7 resident mathematics teachers as well as other teachers who were expected to implement the MProSE lessons in the next round. A typical team meeting had four segments. Each session usually began with a review of the previous lesson conducted. Next, a team member would share the presentation slides and material for the next lesson that he or she prepared based on the resources provided by the NIE researchers.

---

[1] In Singapore, secondary students are channeled into three ability streams – Express, Normal (Academic), and Normal (Technical).

The team would then typically spend time solving the Problem of the Day (PoD) as well as the problem for the next homework (HW).

Through these sharing sessions, team members helped each other in mathematical problem solving and, more importantly, developed a shared understanding of the learning focus. After most meetings, a short email would be drafted to record key reflections and decisions. The emails also served to inform the NIE researchers of the school's progress. It also served as a means to seek advice and assistance from the researchers when necessary. For example, there was an instance when the team found the PoD proposed by the MProSE guide to be unsuitable for the students as its solution required knowledge of factorisation which the students had yet to learn. The NIE researchers were available to provide guidance on adaptations that could be made. (See Appendix E for a comparison of the PoD and HW proposed in the MProSE programme and the actual ones implemented in Bedok South Secondary School.)

As the lessons progressed, the team also found it useful to have a member of the team conduct the next lesson before the other classes while the other members of the team observed. This lesson came to be known as the 'demonstration lesson'. Immediately after the demonstration lesson, adaptations were made to the material based on the observers' inputs sometimes resulting in a complete overhaul of the lesson plan and focus. For example, during Lesson 5, the team recognised that the students needed more time to engage with the problems provided. As such, the lesson focus for the other two classes was changed and another session was added. A make-up session was conducted for the demonstration class to cover the gaps that were identified.

The amount of preparation and the commitment demanded from the teachers to adopt a curricular innovation are not to be underestimated. We personally experienced the challenge of professional commitment needed to adapt a curricular innovation that required significant changes on our part. The first step to adoption is the mental 'buy-in' that the changes would indeed benefit the students. Additionally, we had to be willing to devote time to attend the training workshop to enhance our capacity to teach problem solving, and to participate in sharing sessions to build a shared understanding of the MProSE design implementation. And, finally, we had to be creative and flexible enough in planning and designing demonstration lessons from which the other teachers can also learn to teach the problem solving lessons to their best ability.

Comments from the NIE Researcher.

On the part of the researchers, they had to continually support the research users in their attempts to put theory into practice, especially considering that teachers may not be well-positioned to access and modify research findings easily. It was also clear from the users' recollection that a special researcher-practitioner relationship is crucial to sustain the adoption process.

A typical lesson. The MProSE experience for our students typically started with a review of the previous session's HW problem. This usually involved some students sharing their solutions with the class, sparking off discussion of the problem situation. More often than not, teachers would discover that there is a wide range in the quality of students' responses. A few students showed almost complete solutions, while others showed little effort in their attempts. There was also a wide range in the quality of their writing, ranging from well-organised prose to scribbling. While some of the students may have been disinterested, there were others who were motivated but found it hard to express their thoughts on the Practical Worksheet.

The class would then be led to a teachers' exposition on the focus of the day. These foci included Pólya's four stages, heuristics, as well as Schoenfeld's conception of metacognition and resources, as detailed in the MProSE guidebook for mainstreams schools.

A PoD would then be introduced to allow pupils to make sense of these concepts in context. Students would work in pairs to solve the PoD and a discussion would ensue. While the students were working on the problem, both the teacher and an assistant would move around the class to help students in need. We provided scaffolding for students' attempts at the problem by giving increasingly specific hints as described in the

MProSE guidebook. Table 1 describes these levels and provides examples of specific scaffolding questions based on the Lockers Problem. (Refer to Appendix B for the text of the problem.)

Table 1

Levels of MProSE scaffolding (from Toh, Quek, Leong, Dindyal, & Tay., 2011, p.22)

Level	Feature	Examples based on the Lockers Problem
0	Emphasis on Pólya stages and control	What Pólya stage are you in now? Do you understand the problem? What exactly are you doing? Why are you doing that?
1	Specific heuristics	Why don't you try with fewer lockers (*use smaller numbers*)? Try *looking for a pattern*.
2	Problem specific hints	Think in terms of the locker rather than the student – what numbers get to touch the locker?

We often found it challenging to decide when to move on to the next pair of students, or when to step in to offer advice. We felt pressured to instruct students on what to do specifically rather than to facilitate their thinking process by asking questions, so that we could move on to the next pair of students. The session would then end by assigning a HW problem that was specially crafted to reinforce those aspects of problem solving highlighted during the lesson.

There were also times during the programme when we felt it necessary to make changes to the original MProSE design to revisit what was recently taught. For example, in Lesson 6, the team decided to revise Pólya's model and on how the Practical Worksheet could be better used to guide and record the students' problem solving efforts. Hence, the MProSE design for Bedok South Secondary School utilised 15 of the original 18 problems. (See Appendix E for a comparison of the list of problems in the original design and the implemented programme.)

Comments from the NIE Researcher.

A lesson well planned is an important precursor of an effective lesson. That, however, is premised upon an enactment of the lesson plan the success of which can sometimes even elude an experienced teacher. The teaching of mathematical problem solving is no easy task, as revealed by decades of research into it. The pedagogical challenges can be formidable in themselves, as attested to by the experiences of the teachers of Bedok South Secondary School. What more if we pile on the mathematical demands upon the pedagogical! As such, for the MProSE design to take root in the classroom, it is crucial that the teachers continue to solve mathematics problems the MProSE way so that they can empathise with the students, for example, in being able to guess well what is in the student's mind as he or she mulls over in devising a plan. Also of importance is their understanding of feelings and emotions (the affective aspect) in problem solving. As the teachers grow in their experience of solving mathematical problems so will they begin to automatise the 'next scaffolding move' to help the students.

Assessment. Assessment of students' learning took the form of a portfolio as well as a paper-and-pencil test at the end of the ten sessions. For the portfolio, we did not require the students to submit their attempts of all the PoDs and HWs. Instead, the teacher specified one PoD and one HW to be submitted, and allowed the students to choose one more PoD and one more HW which they considered represented their best effort.

The portfolio and test were graded according to the rubric suggested by the NIE researchers. (See Appendix C for the marking rubric.) The teachers met for a benchmarking session after marking common samples of good and poor quality work. To improve students' motivation to engage fully in the programme, the students were informed that their performance in the MProSE module contributed a small portion to their Continuous Assessment marks for the year.

Comments from the NIE Researcher.

The assessment practice was not only crafted to motivate student engagement in problem solving. It also aimed to convey to students what is valued in a mathematics class. The assessment rubric serves to focus the students' attention on the crucial aspects of problem solving and inculcate in them the mental discipline of working through Pólya's stages and the development of metacognitive control.

---

3 Reflections

At the end of the MProSE implementation, the Bedok South Secondary team had several meetings to collate feedback and review aspects of the programme. We share some of our reflections and elaborate on how they might influence our plans for future implementation.

Choice of students. For the current run of the programme, we chose to work with our Year 7 Express stream students. Although we found that the problems provided might have been more suitable for use with Year 8 students, the team believes that it may be better to mould students' attitudes and practices right from the beginning of their secondary school life. We also believe that the students need time to develop the intended skills. Hence, working with Year 7 students gives us an opportunity to continue the programme for the same group of students over the next four years so that we can extend and refine their problem solving abilities and mathematical communication skills. This suggests that we would need to customise the programme materials for Year 7 pupils to take into consideration their prior knowledge and the new ideas to be learnt in the current syllabus.

Teacher preparation. We felt that while the teacher preparation workshop allowed us to engage in mathematical problem solving and to go through aspects of students' learning experiences, we would have preferred more exposure to the process of facilitating mathematical problem solving by observing others in action. As one teacher said, "the art of facilitating problem solving comes with experience and observation of others doing so. As a beginning teacher with limited experience, I had some trouble ascertaining the right time to prompt with a question, give a hint or reveal the answer. I found that observing my colleague's lesson really helped me pick up some strategies that I could use with my students." Hence, having peer observations as part of teacher preparation might have been helpful and could have been done from the start of the programme. It might also have been good if we had the opportunity to observe more knowledgeable others in action, too.

Some teachers expressed a desire to attend courses to learn more effective questioning techniques. While the MProSE guidebook provided some ideas on the focus and expectations for each session, the team found that the effort spent on discussing the lesson plan in detail, even to the extent of spelling out the questions to ask and how to lead students, helped us to better anticipate students' responses and how to scaffold their learning experiences. However, due to busy schedules, we were not able to do so for all the lessons, especially the first few ones. It might have been better if we had been able to devote more time for teachers to refine their lesson plans. A more formal template to guide our discussions could have also been helpful.

Teachers could also have been better prepared to appreciate the criteria for selecting the problems used, as well as to appreciate the appropriate level of rigour and depth in reasoning that should be expected from the students in the problems selected at the different stages. We had difficulty understanding why for some problems there was a need for rigorous proofs, while for others, it was sufficient to address the particular cases.

In our attempts to modify the problems to engage our students, we might also have missed out on important considerations such as the opportunities afforded by the problems to use particular heuristics. A case in point was our attempt to modify HW 8. We replaced the proposed problem because we were concerned that the concepts involved in solving it were still inaccessible to our students. However, the problem we replaced it with did not offer the students as much opportunity to make use of certain heuristics compared to the original problem.

Schedule and duration of the lessons. The schedule was a concern for the students. Some students did not like the programme as it was conducted as a supplementary lesson outside of the regular school hours once a week. Already drained from a whole day of lessons, students found it a burden to work on the challenging questions in the afternoons. They grudgingly attended lessons which they thought were enrichment classes. However, they felt that the classes were irrelevant as the content taught would not be tested in the O-level examination. To solve this problem, we suggest that the lessons be carried out in normal curriculum hours during the next implementation. We are also thinking of including a more challenging question in their end-of-year examination. This will test the students' problem solving skills as well as provide them with a better sense of purpose for learning these skills.

We also found the time allocated for each lesson to be too short. During this implementation, one hour a week was set aside for MProSE lessons. Often there was not enough time to do all these: introduce the concepts required in the module, allow students time to explore and solve the problem, and to go through the solutions in class. We thought that a longer session – perhaps stretching to 90 minutes – might be better to provide opportunities for exploration and engagement with the PoDs and to review the previous HW.

To help highlight the relevance of MProSE to the teaching of standard content, we also want to explore aligning the PoDs and HWs to the regular lessons and to provide time within the curriculum for students to engage in these learning experiences. This will also make it possible for students to engage in MProSE over a longer period of time.

Practical Worksheet. In the beginning, the students found the Practical Worksheet very difficult to use. After the first two lessons, students still left many parts unanswered. Thus, the team came together to redesign the layout of the worksheet.

For the initial lessons, we attempted to reduce the number of pages from four to two, providing specific spaces for each of the guiding questions. The smaller number of pages made the worksheet look less intimidating to the students while they were guided visually to answer each question step by step. Teachers also found it easier to identify the guiding questions that students had yet to address. As the lessons progressed, we found that Pólya's first and fourth stages required more space, resulting in further modifications. (See Appendix A.2 for one of the modified versions of the Practical Worksheet used in Bedok South Secondary School.)

Despite the attempts to make the worksheets easier to use, students still struggled to pen their thoughts on paper for the most part of the course. The situation improved slightly when teachers extended what was originally planned as Lesson 5 to two lesson durations (i.e., Lesson 5 and 6). Once again, they reminded students of the rationale of the course and modelled the use of the Practical Worksheet in detail. While we did not find much difference in the quality of students' responses after the modification of the Practical Worksheets, our experience suggested that it might be useful to customise worksheets according to the focus of the day such that students' learning goals of the specific lesson are reinforced. Customised worksheets can also help teachers identify and make sense of students' attempts at problem solving.

Lesson design. It was good that students had opportunities to practise the use of various heuristics. Some students had already been exposed to using heuristics in primary school. Thus, this course helped reinforce and strengthen their use as they could hardly do so in their regular mathematics lessons. However, there were a number of heuristics that were new to them. The limited time spent in learning these new heuristics in the programme might not have allowed students to gain sufficient familiarity or fluency to use of these heuristics successfully. Thus, we think that future implementations should involve more explicit instruction of these new heuristics before expecting students to use them appropriately in solving problems.

Students started to lose confidence in using Pólya's four stages by the third lesson as they could not experience successes. The hour-long sessions were not long enough for our students to attempt and solve the problems. Thus, teachers ended up giving students the answers so that they would have some knowledge to attempt their HW problem. Even then students still struggled with their homework. Since some of the questions were also beyond the ability of our students in terms of content knowledge, we sourced for other questions from the pool of problems supplied by the NIE researchers and also reordered the questions so that students would feel a sense of achievement by being able to solve the questions by themselves. We also wanted them to see that the questions could be related to the mathematics syllabus they were studying as some students did not see the relevance of studying problem solving.

As we gain greater awareness of the problem solving process and the demands of facilitating students' learning, we hope to incorporate more pedagogical strategies that will support the students' problem solving processes better. For instance, we can employ purposeful grouping of students in such a way that the motivated and able students can serve as resources that teachers can tap upon. Differentiated task experiences can also be designed for students with different abilities.

Assessments and rubrics. One of the challenges in assessment was to award marks for alternative solutions. As it is, most of the problems posed to the students were relatively challenging for them. To have them provide an adequate solution was an achievement. Thus, students found the need to seek alternative solutions extremely challenging. This was especially true during the timed final assessment. Practically all the students could not obtain the two marks that were allotted for finding alternative solutions in the marking rubric. On the other hand, it was important to highlight this aspect of mathematical problem solving. Hence, further refinement to the rubrics or the problem used for the Practical Test would be necessary. The original rubric could still be used for evaluating the PoDs and HWs.

Marking the students' answers on the Practical Worksheet was also challenging. It was time consuming, and there were ambiguous answers that were still difficult to mark despite the rubric. While the benchmarking exercise helped, especially in evaluating the test and the two teacher-selected problems, the other two student-selected questions took a while to go through. Some adaptations to the rubric may be necessary for the markers to evaluate the students' work more efficiently.

Impact on students. By the end of the programme, after we marked the final test question, we could see that more students were motivated to go beyond solving the problem and to go on to Stage 4 of expanding the problem. In doing so, they were trying to delve deeper into the concepts to create authentic questions. Through the course, teachers also noticed that students were becoming more persevering. It was also heartening to see that there were a handful of students in every class who were more enthusiastic in the MProSE lessons than in their regular mathematics lessons. These students were also scoring better in their MProSE assignments than in their regular class assignments and tests. They also seemed to looked forward to the PoDs and HW questions.

As lessons were scaffolded by the teachers, a number of students gradually got better at using the Practical Worksheet and got into the rhythm of going through Pólya's problem solving stages by the second half of the course. In the last few problems and the Practical Test that the students worked on, it was evident that students went through Pólya's stages with appropriate looping back to certain stages as reflected by their outputs in their worksheets

Impact on teachers. Over the duration of the implementation of the programme, we took to redesigning parts of the Practical Worksheet to make it more accessible to students. The first version we made was designed to scaffold students' processes by judiciously allotting the space available for each stage. We also reduced the number of pages so that the worksheet looked less daunting to use. However, as the lessons proceeded, we realised that the amount of space we had provided for the different segments was inappropriate. For example, after the initial lessons, we found that more space was needed for Stage 2, as the students needed more space to explore so that they could better appreciate the requirements of the problem. Towards the end of the programme, we also found that the space that we had provided for Stage 4 was still insufficient; hence, further changes were made.

Looking back, these adaptations suggest that the form of the Practical Worksheet could morph with the focus of the lessons. They also reflected our own growing appreciation of the problem solving process.

Participating in the programme has planted in us a desire to enhance our own subject matter and pedagogical knowledge. For instance, we feel that we will benefit from participating in training in specific areas such as questioning techniques. We would also like to learn more about MProSE design principles and the criteria for problem selection.

4 Conclusion

The implementation of the MProSE programme was a challenging yet rewarding experience. While we think that the positive impact on students was not always plain to see over a short period of time, the positive impact on teachers was generally more evident. With greater familiarity with the content and focus of the programme and the associated pedagogy, as well as possible tweaking of the resources, we believe that we will be more effective in facilitating the learning experiences of students and thus achieve better results.

We came into the programme knowing that it would take several iterations for us to be able to carry out the programme as effectively as possible. This first attempt gave us much insight into how we can better facilitate our pupils' learning and enhance our own capacity. It has also given us the confidence to embark on collecting more objective data to guide our future reflections and implementation.

Comments from the NIE Researcher.

The MProSE design has the potential to bring mathematical problem solving to all students, not just the mathematically precocious. The practical value of MProSE is in bringing to teachers usable knowledge of mathematical problem solving and school curricular policy-making to help teachers teach problem solving. The immense diversity of school life, however, and the great welter of factors underlying any teaching and learning of problem solving make it necessary to fine-tune the design while remaining true to the spirit of MProSE so that the design is still valuable to the school. Where the research users do not see the potential advantage of adopting a curricular innovative, there is no reason for them to embrace the innovation.

Curricular programmes that work well with one school under certain conditions may be less effective with another school where the circumstances are different. What practices must schools give up or what adjustments must they make?

Here at Bedok South Secondary School, the teachers rightly deserve more time and support to implement the MProSE design. They should be given time to develop the language to talk about problem solving and its teaching, and to build up the repertoire of classroom moves that they can readily use to respond positively to students' concerns at the different stages of Pólya's model for problem solving. The teachers themselves require time to be familiar with the use and purpose of the Practical Worksheet. We are beginning to see a convergence of information from the professional sharing sessions which led the teachers to challenge and shape curricular policies and classroom practices (e.g., pedagogy). In the process, they continue to grow in their confidence and competence in mathematical problem solving and its teaching. This knowledge which the teachers have co-constructed will remain with the teachers, embedded within the culture of Bedok South Secondary School.

Chapter 7

Another Step towards Mathematical Problem Solving for Everyone

LEONG Yew Hoong   TOH Tin Lam   TAY Eng Guan   TOH Pee Choon
QUEK Khiok Seng   Jaguthsing DINDYAL   HO Foo Him

Due to the purpose of this chapter as one that provides a kind of wrap-up of this book, it was the last to be written. In fact, there was a significant time lapse of about 9 months between the submission dates of all the other chapter manuscripts and the writing of this chapter. In the intervening period, a number of the participating schools had begun the second implementation of the MProSE module in their respective schools, bringing on board the revisions that they proposed in the earlier chapters.

1 Second Implementation of the MProSE module

We collected extensive video data of most of the lessons and the post-lesson meetings conducted in the second implementation. As the analyses of these data are ongoing, we are only able to give a broad overview at this juncture. Table 1 provides a summary of comparison between the first implementation (conducted in 2012 and reported at length in the earlier chapters) and the second implementation (conducted in 2013). Only three schools are included in this review as we have not obtained current data from the other schools.

Table 1

Comparison between the first and second implementation of the MProSE module

	Jurong Secondary		Bedok South Secondary		Tanjong Katong Girls	
	First	Second	First	Second	First	Second
Year Level involved	Year 7 Express	Year 7 Express	Year 7 Express	Year 7 Express	Year 8	Year 8
Place of module	1 hr added to maths curriculum time	Same as previous year	After-school programme	Within curriculum time	After-school programme	Within curriculum time
Curriculum Planning	1 hr a week meeting before module	Less frequent meetings – more reliant on resources from previous year	1 hr a week meeting before and during module	Advanced planning to refine module structure and lessons	Focus on minor tweaks to MProSE lessons	Numerous discussions on fine-grained lesson details
Mathematics Problems	From MProSE guidebook - wholesale	A few replacements from MProSE guidebook	A few replacements from MProSE guidebook	Further refinements based on 1st experience	From the MProSE guidebook – wholesale	Reduced number of problems
Teachers of the module	Selected teachers – including Department Head	Same as previous year, with 2 additions	Selected teachers – including Department Head and Subject Head	Same as previous year, with 2 additions	All Year 8 teachers and others as 'standby'	Same Year 8 teachers as previous year, with one addition
Teacher development	Teachers observe the lesson conducted by the Department Head – time-table adjustment	Pairing teachers conducting the lessons for the first time with experienced ones	Internal problem solving sessions, observation of lessons halfway through the module	Include MProSE researchers in detailed discussion on lesson-to-lesson development	Consultation of MProSE researchers initiated by school	Include MProSE researchers in a few 'critical' planning meetings
Continual Assessment	Practical Worksheet	Practical Worksheet	Modified Practical Worksheet	Modified Practical Worksheet	Practical Worksheet	Practical Worksheet

Across the three schools, there are significant changes in the second implementation. Apart from the features summarised in Table 1, other changes that are significant to us include the following: In Jurong Secondary School, a core group of teachers – led by the department head – spent a considerable amount of time on the new problems that replaced some of the problems in the MProSE guidebook. They were particularly focused on their own familiarity with the mathematics behind the solutions and the expansions. Detailed workings were produced for these problems and provided as resources for other teachers who were teaching the module; In Bedok South Secondary School, the focus was on building resources that teachers would actually use in the classroom. The bundle of resources includes overhead slides and the modified Practical Worksheets (that finally build towards the MProSE Practical Worksheet) for each lesson. The materials underwent substantial revision from the first implementation to take into consideration the re-arrangement of some problems and the concomitant amended flow in the teaching of problem solving processes within the module; In Tanjong Katong Girls' School, the teachers channeled their efforts on building students' positive experience with the MProSE module. Implements towards this end include spending more time with a particular problem at the beginning of the module with the intention of giving students the opportunity to get a more thorough experience with the problem – the actual "acting it out", followed by conjecturing the solution, then doing the expansion of the problem. Students were also divided into assigned groups and given specific roles with the goal of increasing their active participation during MProSE lessons.

Some changes are, to us (as MProSE researchers and designers), clearly steps forward. Examples of these include the movement of the MProSE module from outside-curriculum to within-curriculum, as was the case with Bedok South Secondary and Tanjong Katong Girls'School. We also think that the careful consideration and replacement of the problems to suit the local conditions of each school is the tweaking that needs to be carried out as part of the overall design experiment. Moreover, the specific emphases of each school brought about distinctive local 'flavours' while keeping to the broad parameters of MProSE. These homegrown innovations help to increase school ownership of the enterprise.

However, with each adjustment to improve a certain aspect of implementation, there is a risk that it would hinder or mask other important features. This is perhaps best described by the teachers from Bedok South, "In our attempts to modify the problems to engage our students, we might also have missed out on important considerations such as the opportunities afforded by the problems to use particular heuristics" (Thaven et al., this volume, p. 100). In other words, the teachers recognised that being users (and not the primary designers) of MProSE, they may not have a good enough grasp of the foundational considerations – both mathematically and theoretically, with respect to mathematical problem solving – underpinning the choice of problems and the flow of the lessons. In their attempts at the local level to chop, add, replace, and re-order, teachers may disrupt the balance among the various components and distort the intended problem solving developmental trajectory embedded in the MProSE lessons. This highlights the importance of continuous teacher development as a major feature of our MProSE design.

2 Teacher Development

Teacher development is a very important piece in the entire MProSE Design Experiment. [For details on how MProSE incorporates teacher preparation in the overall design, the reader may refer to Leong, Dindyal, Toh, Quek, Tay, & Lou (2011)]. Since the efficacy of MProSE can only be realised in the testbed of the classroom, and the quality of classroom instruction is dependent mostly on teacher actions and decisions, it is not an exaggeration to state that the success of MProSE stands or falls on the capacity of the teachers.

For this reason, we do not take the narrow view of teacher preparation being restricted only to the teacher workshops that we conducted prior to MProSE implementation in the schools; rather, we take teacher development to mean a continual engagement with the teachers throughout the course of implementation across a number of years. As explicated in the earlier chapters (and summarised in Table 1), the actual form of teacher development varies across schools, but the principle is one of long term support for building a sustainable community of learning to teach problem solving within each school. Unless there is a culture of teachers engaging actively in mathematical problem solving and regular discourse on teaching about problem solving, the goal of embedding problem solving as a mainstay in the school follows the way of most research projects – lots of teacher activity to meet research objectives within the project duration but dies out soon after the project ends.

We are under no illusion that this vision of teacher development is easily realisable. In fact, the challenges are already conspicuous, as is well expressed by Teo et al., "The ... issue was the teachers' readiness to conduct the rather unconventional problem solving lessons. Teachers attended [the] problem solving professional development workshop ... . However, they were still unsure of how to conduct the lessons" (this volume, p. 81). We think that one major hurdle to scale is the teachers' own disposition towards mathematical problem solving. By "disposition" is meant the overall cognitive and affective orientation – such as whether one buys in to the idea, enjoys engaging in problem solving, is humble to admit deficiencies in certain knowledge domains (and seeks actively to develop relevant knowledge and skills), and actually employs the Pólya's Stages to find a way forward when stuck – towards the MProSE design.

We do not know of a sure-fire method to help teachers overcome these mental challenges, other than to encourage teachers in each school to set aside time in their internal professional development meetings to engage regularly in the work of problem solving in order to experience the efficacy of these processes for themselves. As a form of concrete support, we set up an online platform where suitable mathematics problems are regularly posed to teacher participants across all MProSE schools. Solution processes were posted and discussed after some time lapse to give teachers a temporal window to attack the problems. Apart from serving as an avenue to encourage teachers' practice of regular problem solving, we think it can be a step towards enlarging the professional learning community beyond the boundaries of one school as they interact online with teachers across other schools. In addition to these e-exchanges, we organise symposia on a yearly basis where teachers not only hear the feedback and updates from the MProSE researchers but also learn from the experience of implementation from other participating schools.

Other than developing positive disposition and proficiency in problem solving, we think it is also important that teachers develop effective instructional models for actual classroom practice in the teaching of problem solving. While we do not advocate a one-size-fits-all model when it comes to teaching approaches, nor a simplistic adherence to theoretical ideals such as being purely "constructivist", "conceptually-oriented", or "inquiry-based", we recognise that significant changes in teachers' instructional behaviours and decision-making may be necessary to complement the problem solving goals we intend to achieve in class. Some of the changes needed include the curbing of the instinct of giving immediate answers and shifting towards the skill of scaffolding, the ability to elicit from students the underlying workable problem solving processes instead of direct telling, and the careful timing of various junctures to introduce suitable heuristics to maximise motivation rather than merely following a fixed schedule.

The school-based initiative – such as are currently practiced by Jurong Secondary and Bedok South Secondary – of making provisions for teachers to peer-observe problem solving lessons is a helpful step towards building capacity for teaching problem solving. There is room to develop this informal observation and feedback process into a more organised goal-oriented teacher development programme within the schools' professional learning time slots. However, we recognise that real-time observation of lessons require rather substantial reshuffles in time-tabling and is therefore not feasible as a sustainable practice. In this regard, we can tap on the videos of lessons that we collected as resources for teacher development. We are currently exploring the use of these videos as portraits for teachers' reflections of how the lesson objectives are achieved and also as a visual trigger for discussion into alternative instructional routes to achieve these objectives.

3 MProSE Diffusion

With regards to MProSE, we think of diffusion in at least two levels of grain-sizes in terms of social units: (1) at the within-school level, diffusion refers to the growth in the school's capacity and scope to implement the MProSE design efficaciously; (2) at the level of school systems, we think of diffusion as the process of influencing more schools to adopt the MProSE design. As the latter is beyond the scope of this book (and perhaps the subject of subsequent MProSE-related reports), we focus on the former.

The participating schools are taking steps to enlarge the pool of teachers with the experience to carry out the MProSE lessons. This is clear from Table 1 under the category of "Teachers of the module". It is interesting to note that all the three schools moderate the rate of inducting teachers into the responsibility of teaching the module. Basically the same group of teachers taught during both implementations, with only one or two teachers included in the second round. We think the rate of growth of the teacher pool and the overlap of teachers across implementations are significant factors in determining the success of the diffusion. On one hand, when teachers are not suitably initiated to MProSE – and given the opportunity to undergo the process of teacher development as detailed in the earlier section of this chapter – the buy-in by teachers and also the efficacy of the MProSE lessons can be critically affected to the point that negativism can derail the entire enterprise; On the other hand, too slow diffusion means the whole process takes many years and thus subjected to fatigue and risk of MProSE being supplanted by other priorities due to changes of personnel especially in significant positions within the school.

The goal of within-school diffusion is one where there is stability – culturally, structurally, and knowledge-wise – residing within the mathematics department and its instructional programme to weather the challenges of usual staff changes and educational short-termism. To achieve that, we think these features are necessary: (1) Obtain strong support from school leaders, especially in the first three years of the project; (2) Build MProSE lessons as a module *within* normal curricular hours – a feature that all the three schools currently shares – as soon as possible. This makes reversing – by removing the module – a decision that requires careful deliberation; (3) Place capacity development of all teachers within the mathematics department a top priority; (4) Begin with a core group of teachers that include opinion shapers within the department; (5) Carefully plan a model of moderate spread of capacity with deliberate overlaps of teachers who have experienced teaching the module and are positive about implementation with others who are looking in 'from the outside'; (6) Include teacher problem solving as a regular feature of professional development meetings; (7) Infusion of problem solving into the regular teaching of mathematics.

4 MProSE Infusion

While the problem solving module is intended primarily for teaching *about* mathematical problem solving, our conception of "infusion" includes aspects of teaching mathematics *for* problem solving and teaching mathematics *through* problem solving. The vision is this: Through the problem solving module, the students learn the processes, language, and assessment relating to problem solving; these problem solving heuristics, stages, and skills can then be "infused" into regular mathematics lessons throughout the school levels as they learn particular mathematics content. The junctures where problem solving becomes conspicuous – for example, through the use of the Practical Worksheet – in the classroom can occur at the beginning, at the end, or at other suitable points in the teaching of a specific mathematics topic. When used at the start, it is likely to be a case of teaching mathematics *through* problem solving. For example, instead of direct teaching of the cosine rule, it can be posed as a problem of finding the length of a side of a triangle given that the lengths of the other two sides and their included angle are known. Teaching mathematics *for* problem solving can be used at the end of a module where students use the contents learnt in the module to solve a related problem. A mixture of "for" and "through" can be conceived for problems introduced in the middle of a module. The goal of infusion is that problem solving becomes a mainstay in the mathematics curriculum instead of appearing merely in 'boutique' lessons and remaining separated from typical experience of doing mathematics.

At the point of writing, the participating schools have yet to begin the infusion stage of MProSE. Like the stages before, we anticipate challenges in similar domains – teacher preparation, choice of suitable problems and the suitable junctures where they may be inserted, and quality of students' experience with problem solving. We are excited at the prospect of venturing into a territory where few have succeeded (Stacey, 2005). The story of this next stage is, we hope, the contents of another book in this series.

REFERENCES

Baker, E.L. (2007) Principles for scaling up: Choosing, measuring effects, and promoting the widespread use of educational innovation. In B. Schneider & S.-K. McDonald (Eds.), *Scale-up in education* (pp. 37–54). Lanhm, MD: Rowman & Littlefield.

Ball, D. L. (2000). Bridging practices: Intertwining content and pedagogy in teaching and learning to teach. In J. Bana & A. Chapman (Eds.), *Mathematics education beyond 2000 Proceedings of the 23rd annual conference of the Mathematics Education Research Group of Australasia* (pp. 3–10). Perth: MERGA.

Black, P. (2009). In response to: Alan Schoenfeld. *Educational Designer. 1*(3), 1–6.

Brown, A.L. (1992). Design experiments: Theoretical and methodological challenges in creating complex interventions. *The Journal of the Learning Sciences, 2*, 137–178.

Burrill, G., Allison, J., Breaux, G., Kastberg, S., Leatham, K., & Sanchez, W. (2002). *Handheld graphing technology in secondary mathematics: Research findings and implications for classroom practice.* Dallas, TX: Texas Instruments.

Cobb, P., Confrey, J., diSessa, A., Lehrer, R., & Schauble, L. (2003). Design experiments in educational research. *Educational Researcher, 32*(1), 9–13.

Collins, A. (1999). The changing infrastructure of education research. In E. C. Langemann & L. S. Shulman (Eds.), *Issues in education research* (pp. 15–22), San Francisco, CA: Jossey-Bass.

Doerr, H. & Lesh, R. (2003). Designing research on teachers' knowledge development. In L.Bragg, C. Campbell, G. Herbert, & J. Mousley (Eds.), *Mathematics Education Research: Innovations, Networking, Opportunities* (pp. 262–269). Proceedings of the 26th annual conference of the Mathematics Education Research Group of Australasia, Geelong.

English, L., Lesh, R., & Fennewald, T. (2008). *Future directions and perspectives for problem solving research and curriculum development.* Paper presented at the 11th International Congress in Mathematics Education, Jul 6–13, 2008, Mexico.

Foong, P.Y. (2009). Review of research on mathematical problem solving in Singapore. In K. Y. Wong, P. Y. Lee, B. Kaur, P. Y. Foong & S. F. Ng (Eds.), *Mathematics education: The Singapore journey* (pp. 263–300). Singapore: World Scientific.

Foong, P. Y., Yap S. F., & Koay, P. L. (1996). Teachers' concerns about the revised mathematics curriculum. *The Mathematics Educator, 1*(1), 99–110.

Fuchs, D., & Fuchs, L. (1998). Researchers and teachers working together to adapt instruction for diverse learners. *Learning Disabilities Research and Practice, 13*(3), 162–170.

Gorard, S., with Taylor, C., (2004). *Combining methods in educational research and social research.* Maidenhead, England: Open University Press.

Ho, K. F., & Hedberg, J. G. (2005). Teachers' pedagogies and their impact on students' mathematical problem solving. *Journal of Mathematical Behavior, 24*(3–4), 238–252.

Hogan, D. (2007). Towards "invisible colleges": Conversation, disciplinarity, and pedagogy in Singapore. Slide presentation available from Office of Education Research, National Institute of Education, Nanyang Technological University.

Institute of Educational Sciences (2004). *Reading comprehension and reading scale-up research CFDA84.305G.* Washington DC: Department of Education.

Kaput, J. (1992). Technology and mathematics education. In D. Grouws (Ed.), *A handbook of research on mathematics teaching and learning* (pp. 515–556). New York: MacMillan.

Kaput, J. (1994). Democratizing access to calculus: New routes using old roots. In A. Schoenfeld (Ed.), *Mathematical thinking and problem solving* (pp. 77–155). Hillsdale, NJ: Lawrence Erlbaum Associates.

Kaput, J. (1997). Rethinking calculus: Learning and thinking. *The American Mathematical Monthly, 104*(8), 731–737.

Kaur, B. (2009). Performance of Singapore students in Trends in International Mathematics and Science studies (TIMSS). In K.Y. Wong, P.Y.Lee, Kaur, B., P.Y. Foong, & S.F. Ng (Eds.), *Mathematics education: The Singapore journey* (pp. 439–463). Singapore: World Scientific.

Kilpatrick, J. (1985). A retrospective account of the past 25 years of research on teaching mathematical problem solving. In E. A. Silver (Ed.), *Teaching and learning mathematical problem solving: Multiple research perspectives* (pp. 1–15). Hillsdale, NJ: Lawrence Erlbaum Associates.

Koay, P. L., & Foong, P. Y. (1996). *Do Singaporean pupils apply common sense knowledge in solving realistic mathematics problems?* Paper presented at the Joint Conference of Educational Research Association, Singapore, and Australian Association for Research in Education, November 25–29, 1996, Singapore.

Kuehner, J. P., & Mauch, E. K. (2006). Engineering applications for demonstrating mathematical problem solving methods at the secondary education level. *Teaching Mathematics and its Applications, 25*(4), 189–195.

Lampert, M. (2001). *Teaching problems and the problems of teaching.* New Haven, CT: Yale University Press.

Leikin, R., & Kawass, S. (2005). Planning teaching an unfamiliar mathematics problem: The role of teachers' experience in solving the problem and watching pupils solving it. *Journal of Mathematical Behavior, 24*(3–4), 253–274.

Leong, Y.H., Dindyal, J., Toh, T.L., Quek, K.S., Tay, E.G., & Lou, G.T. (2011). Teacher preparation for a problem-solving curriculum in Singapore. *ZDM: The International Journal on Mathematics Education, 43*(6–7), 819–831.

Leong, Y.W. (2012). The practical worksheet: Scaffolding and assessing the problem solving process. *Unpublished Masters dissertation*, National Institute of Education, Singapore.

Lemke, J. L., & Sabelli, N. H. (2008). Complex systems and educational change: Towards a new research agenda. *Educational Philosophy and Theory, 40*(1), 118–129.

Lesh, R., & Zawojewski, J. S. (2007). Problem solving and modelling. In F. Lester (Ed.), *The Second Handbook of Research on Mathematics Teaching and Learning* (pp. 763–804). Charlotte, NC: Information Age Publishing.

Lester, F. K., & Kehle, P. E. (2003). From problem solving to modelling: The evolution of thinking about research on complex mathematical activity. In R. A. Lesh & H. M. Doerr

(Eds.), *Beyond constructivism: Models and modelling perspectives on mathematical problem solving, learning, and teaching* (pp. 75–88). Mahwah, NJ: Lawrence Erlbaum Associates.

Lester, F. K. (1994). Musing about mathematical problem-solving research: 1970–1994. *Journal of Research in Mathematics Education, 25,* 660–676.

Lim, S.K. (2002). Mathematics education within the formal Singapore education system: Where do we go from here? In *Mathematics Education for a Knowledge-based Era: Proceedings of Second East Asia Regional Conference on Mathematics Education.* Singapore: National Institute of Education.

Lowrie, T., (2012). Visual and spatial reasoning: The changing form of mathematics representation and communication. In B. Kaur & T.L. Toh (Eds.), *Reasoning, communication and connections in mathematics: Yearbook 2012, Association of Mathematics Educators* (pp. 149–168). Singapore: World Scientific.

Middleton, J., Gorard, S., Taylor, C., & Bannan-Ritland, B. (2006). The 'compleat' design experiment: From soup to nuts. *Department of Educational Studies Research Paper 2006/05 University of York.*

Ministry of Education (MOE). (2006). *Secondary mathematics syllabuses,* Singapore: Ministry of Education, Curriculum Planning and Development Division. http://www.moe.gov.sg/education/syllabuses/sciences/files/maths-secondary.pdf

Pólya, G. (1945). *How to solve it.* Princeton: Princeton University Press.

Quek, K.S., Dindyal, J., Toh, T.L., Leong, Y.H., & Tay, E.G. (2011). Problem solving for everyone: a design experiment. *Journal of the Korea Society of Mathematical Education Series D: Research in Mathematical Education, 15*(1), 31–44.

Rogers, E.M. (1983). *Diffusion of Innovations (3rd Ed).* New York: The Free Press.

Rogers, E. M. (2003). *Diffusion of innovations (5th Ed).* New York: Simon and Schuster.

Roschelle, J., Tatar, D., Schechtman, N., & Knudsen, J. (2008). The role of scaling up research in design for and evaluating robustness. *Educational Studies in Mathematics, 68,* 149–170.

Schoenfeld, A. (1985). *Mathematical problem solving.* Orlando, FL: Academic Press.

Schoenfeld, A. (1992). Learning to think mathematically: Problem solving, metacognition, and sense making in mathematics. In D. A. Grouws (Ed.), *Handbook of research on mathematics teaching and learning,* 334–370. New York: Macmillan.

Schoenfeld, Alan H. (1994). *Mathematical thinking and problem solving.* Hillsdale, NJ, England: Lawrence Erlbaum Associates, Inc.

Schoenfeld, A. H. (2007). Problem solving in the United States, 1970–2008: Research and theory, practice and politics. *ZDM The International Journal on Mathematics Education, 39,* 537–551.

Schoenfeld, A. H. (2009). Bridging the cultures of educational research and design. *Educational Designer, 1*(2), 1–22.

Schoenfeld, A. H. (2011). *How we think: A theory of goal-oriented decision making and its educational applications.* New York, NY: Routledge.

Schneider, B., & McDonald, S. (Eds.) (2007). *Scale-up in education: Ideas in principle, vol. 1.* Lanham: Rowman & Littlefield.

Schneider, B., & McDonald, S. (Eds.) (2007). *Scale-up in education: Issues in practice, vol. 2.* Lanham: Rowman & Littlefield.

Schroeder, T. L., & Lester, F. K. (1989). Developing understanding in mathematics via problem solving. In P. R. Trafton & A. P. Shulte (Eds.), *New directions for elementary school mathematics* (pp. 31–42). Reston, Virginia: The National Council of Teachers of Mathematics, Inc.

Silver, E. A., Ghousseini, H., Gosen, D., Charalambous, C., & Strawhun, B. T. F. (2005). Moving from rhetoric to praxis: Issues faced by teachers in having students consider multiple solutions for problems in the mathematics classroom. *Journal of Mathematical Behavior, 24*(3–4), 287–301.

Stacey, K. (2005). The place of problem solving in contemporary mathematics curriculum documents. *Journal of Mathematical Behavior, 24*(3–4), 341–350.

Tan, S. C., Divaharan, S., Tan, L., & Cheah, H.M. (2011). *Self-directed learning with ICT: Theory, practice and assessment.* Singapore: Ministry of Education.

Tay, E. G., Quek, K. S., Dong, F. M., Toh, T. L., & Ho, F. H., (2007) Mathematical problem solving for Integrated Programme students: The Practical paradigm, Proceedings EARCOME 4 2007: Meeting the Challenges of Developing a Quality Mathematics Education Culture, 463–470.

Toh, T.L., Quek, K.S., Leong, Y.H., Dindyal, J., & Tay, E.G. (2011). *Making mathematics practical: An approach to problem solving.* Singapore: World Scientific.

Wood, T. & Berry, B. (2003). What does "design research" offer mathematics education? *Journal of Mathematics Teacher Education, 6,* 195–199.

Wood, T., Cobb, P., & Yackel, E. (1995). Reflections on learning and teaching mathematics in elementary school. In L. P. Steffe & J. Gale (Eds.), *Constructivism in education* (pp.401–422). Hillsdale, NJ: Lawrence Erlbaum Associates.

Yeo, K. K., (2009). Integrating Open-Ended Problems in the Lower Secondary Mathematics Lesson. In B. Kaur, B.H. Yeap & M. Kapur (Eds.), *Mathematical Problem Solving* (pp. 226–240). Singapore: World Scientific.

APPENDICES

Appendix A.1
The Practical Worksheet (Original)

Practical Worksheet

Problem

Instructions

- You may proceed to complete the worksheet doing stages I – IV.
- If you wish, you have 15 minutes to solve the problem without explicitly using Pólya's model. Do your work in the space for Stage III.
  - If you are stuck after 15 minutes, use Pólya's model and complete all the stages I – IV.
  - If you can solve the problem, you must proceed to do stage IV – Check and Expand.

I Understand the problem

(You may have to return to this section a few times. Number each attempt to understand the problem accordingly as Attempt 1, Attempt 2, etc.)

    (a) Write down your feelings about the problem. Does it bore you? scare you? challenge you?

    (b) Write down the parts you do not understand now or that you misunderstood in your previous attempt.

    (c) Write down your attempt to understand the problem; and state the heuristics you used.

<u>Attempt 1</u>

Page 1

II Devise a plan

(You may have to return to this section a few times. Number each new plan accordingly as Plan 1, Plan 2, etc.)
  (a) Write down the key concepts that might be involved in solving the problem.
  (b) Do you think you have the required resources to implement the plan?
  (c) Write out each plan concisely and clearly.

Plan 1

III Carry out the plan

(You may have to return to this section a few times. Number each implementation accordingly as Plan 1, Plan 2, etc., or even Plan 1.1, Plan 1.2, etc. if there are two or more attempts using Plan 1.)

(i) Write down in the *Control* column, the key points where you make a decision or observation, for e.g., go back to check, try something else, look for resources, or totally abandon the plan.

(ii) Write out each implementation in detail under the *Detailed Mathematical Steps* column.

Detailed Mathematical Steps	*Control*
<u>Attempt 1</u>	
	Page 3

IV Check and Expand

(a) Write down how you checked your solution.

(b) Write down your level of satisfaction with your solution. Write down a sketch of any alternative solution(s) that you can think of.

(c) Give one or two adaptations, extensions or generalisations of the problem. Explain succinctly whether your solution structure will work on them.

Appendix A.2
The Practical Worksheet
(Modified Design by Bedok South Secondary School)

Mathematics Practical Worksheet

Session: **POD/HW**

Instructions

1. You may proceed to complete the worksheet doing stages I – IV.
2. If you wish, you have 15 minutes to solve the problem without explicitly using Polya's model. Do your work in the space for Stage III.
· If you are stuck after 15 minutes, use Polya's model and complete all the stages I – IV.
· If you can solve the problem, you must proceed to do stage IV – Check and Expand
· You may have to return to this section a few times. Number each attempt to understand the problem accordingly as Attempt 1, Attempt 2, etc.

Problem Of the Day :

Stage I: Understand the Problem

(a) Write down your feelings about the problem.

(b) Write down the parts you do not understand now or that you misunderstood in your previous attempt.

(c) How are you trying to understand the problem? State the heuristics you used.

Stage II: Devise a Plan

(a) Write down the key concepts that might be involved in solving the problem.
(b) Do you think you have the required resources to implement the plan?

(c) Write out each plan concisely and clearly.

Stage III: Carry Out The Plan

(i) Write down in the *Control* column, the key points where you make a decision or observation, for e.g., go back to check, try something else, look for resources, or totally abandon the plan.

(ii) Write out each implementation in detail under the *Detailed Mathematical Steps* column.

Detailed Mathematical Steps	Control

Stage IV: Check and Expand

(a) Write down how you checked your solution.

(b) Write down your level of satisfaction with your solution. Write down a sketch of any alternative solution(s) that you can think of.

(c) Give one or two adaptations, extensions or generalisations of the problem. Explain succinctly whether your solution structure will work on them.

Appendix B
Problems Used in the MProSE Guidebook
for Mainstream Schools

Lesson 1
What is a
Problem?
	PoD1: Jug problem
	
	You are given two jugs, one holds 5 litres of water when full and the other holds 3 litres of water when full. There are no markings on either jug and the cross-section of each jug is not uniform. Show how to measure out exactly 4 litres of water from a fountain. Consider the following situations as well.
	
	   i. Get 2 litres from 3 litre and 7 litre jugs.
	   ii. Get 6 litres from 12 litre and 16 litre jugs.
	   iii. Get 12 litres from 18 litre and 24 litre jugs.

	HW1: More on the jug problem Pose 3 questions based on the Jug problem and solve them.
Lesson 2 Polya's Problem Solving Strategy	PoD2: What a sum! Simplify $\dfrac{1}{1\times2}+\dfrac{1}{2\times3}+\dfrac{1}{3\times4}+...+\dfrac{1}{100\times101}$, leaving your answer in the simplest form in terms of n. Justify your answer.
	HW2: What an odd sum! Find the sum of the positive odd numbers $1 + 3 + 5 + 7 + \ldots\ldots + 2011$. Justify your answer.
Lesson 3 Using Heuristics to Understand the Problem	PoD3: Phoney Russian roulette Two bullets are placed in two consecutive chambers of a 6-chamber revolver. The cylinder is then spun. Two persons play a safe version of Russian Roulette. The first points the gun at his hand phone and pulls the trigger. The shot is blank. Suppose you are the second person and it is now your turn to point the gun at your hand phone and pull the trigger. Should you pull the trigger or spin the cylinder another time before pulling the trigger?
	HW3: Want to place your bet? Two stacks of cards are numbered from 1 to 100. You are invited to play the game with the following rule: draw one card from each stack. You win the game if the product of the numbers on the card is 5000 or more. Otherwise you lose the game. Would you want to play the game? Justify your answer.
Lesson 4 Using Heuristics for a Plan	PoD4: Going for a movie On a Tuesday, Alice, Bernice, Carol and Dory, met for a movie. After the movie, they made plans for the next gathering. Alice, Bernice and Carol said that they could only go to the movies every 6, 3 and 4 days respectively, starting from that Tuesday. Dory said that she could go to the movies every day except on Fridays.

	After how many days would the four friends be able to meet again for a movie?
	HW4: Pay rise In 2009, Peter was given a pay rise of 5%. In 2010, he was also given a pay rise of 5%. Is his total pay rise 10% over the two years? Justify your answer.
Lesson 5 The Mathematics Practical	**PoD5: The lockers problem** The new school has exactly 343 lockers numbered 1 to 343, and exactly 343 students. On the first day of school, the students meet outside the building and agree on the following plan. The first student will enter the school and open all the lockers. The second student will then enter the school and close every locker with an even number. The third student will then 'reverse' every third locker; i.e. if the locker is closed, he will open it, and if the locker is open, he will close it. The fourth student will reverse every fourth locker, and so on until all 343 students in turn have entered the building and reversed the relevant lockers. Which lockers will finally remain open?
	HW5: Number of squares The figure is a 7×7 array where each cell is a square. Find the number of squares contained in this 7×7 array.

Lesson 6
 	
 Check and Expand
	PoD6: Let's draw a graph
	
	 In a carpark, it was advertised as such:
	
	   First hour parking: Free
	
	   Subsequent parking: $1 per hour or part thereof
	
	 Sketch a graph for parking fee versus the number of hours of parking with axes as shown below.
	
	 Parking fees in $
	
	 Total number of hours

	HW6: Let's draw a graph Complete the 4th stage in the Practical Worksheet for the 'Let's Draw a Graph' Problem by posing as many problems as possible.

Lesson 7 More on Adapting, Extending and Generalising	PoD7: More about rational numbers Is it true that the sum of two rational numbers is always rational? Justify your answer.
	HW7: Circle problem Can you construct a circle where both the circumference and the diameter are integer lengths? Justify your answer.
Lesson 8 Schoenfeld's Framework	PoD8: Averages a) A boy claims that when he left school X and joined school Y, he raised the average IQ of both schools. Explain if this is possible. b) A striker in football is rated according to the average number of goals he scores in a game. Wayne had a higher average than Carlos for games in the year 2007. He also had a higher average than Carlos for games in the year 2008. Can we say that Wayne is a better striker than Carlos over the years 2007-2008?
	HW8: Intersection of Two Squares Two squares, each s on a side, are placed such that the corner of one square lies on the centre of the other. Describe, in terms of s, the range of possible areas representing the intersections of the two squares.
Lesson 9 More on Control	PoD9: Pattern is not proof A sequence of numbers u_1, u_2, u_3, u_4, ..., is defined by the formula $u_n = 2n - 1 + (n-1)(n-2)(n-3)(n-4)(n-5)$, where $n = 1, 2, 3, 4, \ldots$ Find u_1, u_2, u_3, u_4, u_5 and u_6.
	HW9: Last digit Find the last digit of 13^{77}.
Lesson 10 Revision	PoD10: Nice numbers A 'nice' number is a number that can be expressed as the sum of a string of two or more consecutive positive integers. Determine which of the numbers from 50 to 70 inclusive are nice.

Appendix C
Rubric for Assessing Problem Solving

Name: _____

Pólya's Stages		Marks
	Descriptors/Criteria *(evidence suggested/indicated on practical sheet or observed by teacher)*	**Marks**
Correct Solution		
Level 3	Evidence of complete use of Pólya's stages – UP + DP + CP; and when necessary, appropriate loops. [10 marks]	
Level 2	Evidence of trying to understand the problem and having a clear plan – UP + DP + CP. [9 marks]	
Level 1	No evidence of attempt to use Pólya's stages. [8 marks]	
Partially Correct Solution (solve significant part of the problem or lacking rigour)		
Level 3	Evidence of complete use of Pólya's stages – UP + DP + CP; and when necessary, appropriate loops. [8 marks]	
Level 2	Evidence of trying to understand the problem and having a clear plan – UP + DP + CP. [7 marks]	
Level 1	No evidence of attempt to use Pólya's stages. [6 marks]	
Incorrect Solution		
Level 3	Evidence of complete use of Pólya's stages – UP + DP + CP; and when necessary, appropriate loops. [6 marks]	
Level 2	Evidence of trying to understand the problem and having a clear plan – UP + DP + CP. [5 marks]	
Level 1	No evidence of attempt to use Pólya's stages. [0 marks]	

Heuristics		
	Descriptors/Criteria *(evidence suggested/indicated on practical sheet or observed by teacher)*	**Marks**
Correct Solution		
Level 2	Evidence of appropriate use of heuristics. [4 marks]	
Level 1	No evidence of heuristics used. [3 marks]	
Partially Correct Solution (solve significant part of the problem or lacking rigour)		
Level 2	Evidence of appropriate use of heuristics. [3 marks]	
Level 1	No evidence of heuristics used. [2 marks]	
Incorrect Solution		
Level 2	Evidence of appropriate use of heuristics. [2 marks]	
Level 1	No evidence of heuristics used. [0 marks]	
Checking and Expanding		
	Descriptors/Criteria *(evidence suggested/indicated on practical sheet or observed by teacher)*	**Marks**
Checking		
Level 2	Checking done – mistakes identified and correction attempted by cycling back to UP, DP, or CP, until solution is reached. [1 mark]	
Level 1	No checking, or solution contains errors. [0 marks]	
Alternative Solutions		
Level 3	Two or more correct alternative solutions. [2 marks]	
Level 2	One correct alternative solution. [1 mark]	
Level 1	No alternative solution. [0 marks]	

Extending, Adapting & Generalizing		
Level 4	More than one related problem with suggestions of correct solution methods/strategies; or one **significant** related problem, with suggestion of correct solution method/strategy; or one **significant** related problem, with explanation why method of solution for original problem cannot be used. [3 marks]	
Level 3	One related problem with suggestion of correct solution method/strategy. [2 marks]	
Level 2	One related problem given but without suggestion of correct solution method/strategy. [1 mark]	
Level 1	None provided [0 marks]	

Hints given:

Marks deducted: _____

Total marks: _____

Appendix D
On-line Survey Administered to Students
at Tanjong Katong Girls' School

PART I. Please rate each of the statements below using the scale:

Totally Disagree = 1 2 3 4 5 6 = Totally Agree

Also, for each statement, answer **Yes** or **No** to the question: **Was this your view before you attended the module?**

	Statement	Please circle your choice	
		Rating	Was this your view BEFORE the module?
1	Solving mathematics problems requires the use of heuristics.	1 2 3 4 5 6	Yes / No
2	There is more than one method to solve a mathematics problem if I think hard enough	1 2 3 4 5 6	Yes / No
3	All that is needed to solve mathematics problems is repeated practice on solving similar problems.	1 2 3 4 5 6	Yes / No
4	The mathematics teacher must discuss different solutions to a problem with the students.	1 2 3 4 5 6	Yes / No
5	Students who take a long time to solve the problems are not good in mathematical problem solving.	1 2 3 4 5 6	Yes / No
6	Discussion with my friends helps me to solve mathematics problems.	1 2 3 4 5 6	Yes / No
7	It is important for me to check my solution to a problem.	1 2 3 4 5 6	Yes / No
8	Posing related problems (by extending, adapting or generalizing) is a part of learning mathematics.	1 2 3 4 5 6	Yes / No
9	To be successful in solving a problem, I have to monitor my thinking while trying to solve it.	1 2 3 4 5 6	Yes / No

10	Pólya's model of problem solving is useful to me.	1 2 3 4 5 6	Yes / No
11	The practical Worksheet guides me in applying Pólya's model of problem solving	1 2 3 4 5 6	Yes / No
12	The assessment sheets evaluate fairly my problem solving ability.	1 2 3 4 5 6	Yes / No

PART II. For this part of the questionnaire survey, please circle your response to the statement using the scale:

Totally Disagree = 1 2 3 4 5 6 = Totally Agree

1	This module has taught me how to solve mathematics problems.	1 2 3 4 5 6
2	This module was interesting to me.	1 2 3 4 5 6
3	I did not enjoy solving the problems presented in this module.	1 2 3 4 5 6
4	There was sufficient time during class to solve the given problem.	1 2 3 4 5 6
5	The way of thinking taught in this module will help me in other mathematics modules.	1 2 3 4 5 6
6	There should be more than ten hours for this module.	1 2 3 4 5 6

We would also like to seek your views on the following questions.

1. Which is the best part of the module?

2. Which is the worst part of the module?

3. Is there something else that you would like to tell us about the module?

4. What do you think is essential to solve mathematics problems successfully?

Appendix E
The MProSE Problems versus the Implemented Problems at Bedok South Secondary School

Lesson #	Problems proposed by the MPROSE Project Team	Problems used in Bedok South Secondary School
Lesson 1	PoD1: Jug problem	PoD1: Jug problem
	HW1: More on the jug problem	HW1: More on the jug problem
Lesson 2	PoD2: What a sum!	PoD2: Last digit
	HW2: What an odd sum!	HW2: Last digit
Lesson 3	PoD3: Phoney Russian roulette	PoD3: Phoney Russian roulette
	HW3: Want to place your bet?	HW3: Want to place your bet?
Lesson 4	PoD4: Going for a movie	PoD4: Going for a movie
	HW4: Pay rise	HW4: Pay rise
Lesson 5	PoD5: The lockers problem	PoD5: The lockers problem
	HW5: Number of squares	HW5: Number of squares
Lesson 6	PoD6: Let's draw a graph	
Lesson 7	PoD7: More about rational numbers	PoD7: More about rational numbers
	HW7: Circle problem	HW7: Circle problem
Lesson 8	PoD8: Averages	PoD8: Averages
	HW8: Intersection of two squares	HW8: Average speed
Lesson 9	PoD9: Pattern is not proof	PoD9: Pattern is not proof
	HW9: Last digit	HW9: Sum of numbers
Lesson 10	PoD10: Nice numbers	PoD10: Nice numbers

Making Mathematics More Practical

Appendix F
List of Publications for MProSE

Practical Handbook for Mathematics Teachers

Toh, T.L., Quek, K.S., Leong, Y.H., Dindyal, J., & Tay, E.G. (2011). *Making Mathematics Practical: An approach to problem solving.* Singapore: World Scientific.

Chapter in Academic Book

Toh, T.L., Quek, K.S., Leong, Y.H., Dindyal, J., & Tay, E.G. (2011). Assessing problem solving in the mathematics curriculum: A new approach. In Wong, K.Y., Kaur, B. (Eds.), *AME Yearbook 2011: Assessment* (pp. 1–35). Singapore: World Scientific.

Toh, T.L., Quek, K.S., Tay, E.G., Leong, Y.H., & Dindyal, J. (2011). Enacting a Problem Solving Curriculum in a Singapore School. In Bragg, L.A. (Ed.) *Maths is Multi-Dimensional* (pp. 77–86). Melbourne, Australia: Mathematical Association of Victoria.

International Refereed Journals

Dindyal, J., Tay, E. G., Toh, T. L., Leong, Y. H., & Quek, K. S. (2012). Mathematical Problem Solving for Everyone: A new beginning. *The Mathematics Educator, 13,* 51–70.

Quek, K.S., Dindyal, J., Toh, T.L., Leong, Y.H., & Tay, E.G. (2011). Problem solving for everyone: A design experiment. *Journal of the Korea Society of Mathematical Education Series D: Research in Mathematical Education, 15*(1), 31–44.

Tay, E.G., Quek, K.S., Dindyal, J., Leong, Y.H., & Toh, T.L. (2011). Teachers solving mathematics problems: Lessons from their learning journeys. *Journal of the Korean Society of Mathematical Education Series D: Research in Mathematical Education, 15*(2), 159–179.

Leong, Y.H., Toh, T.L., Tay, E.G., Quek, K.S., & Dindyal, J. (2012). Relooking "Look Back": A student's attempt at problem solving using Pólya's model. *International Journal of Mathematical Education in Science and Technology, 43*(3), 357–369.

Leong, Y.H., Tay, E.G., Toh, T.L., Quek, K.S., & Dindyal, J. (2011). Reviving Polya's "Look Back" in a Singapore school. *Journal of Mathematical Behavior, 30*(3), 181–193.

Leong, Y.H., Dindyal, J., Toh, T.L., Quek, K.S., Tay, E.G, & Lou, S.T. (2011). Teacher preparation for a problem solving curriculum in Singapore. *ZDM: The International Journal on Mathematics Education, 43*(6–7), 819–831.

Papers published in proceedings of international conferences

Toh, P.C., Toh, T.L., Ho, F.H., Quek, K.S. (2012). Use of Practical Worksheet in teacher rducation at the undergraduate and postgraduate levels. In J. Dindyal, L.P. Cheng & S.F. Ng (Eds.), *Mathematics education: Expanding horizons, (Proceedings of the 35th annual conference of the Mathematics Education Research Group of Australasia, eBook* (pp. 736–743). Singapore: MERGA

Quek, K.S., Leong, Y.H., Tay, E.G., Toh, T.L., & Dindyal, J. (2012). Diffusion of the mathematics practical paradigm in the teaching of problem solving: Theory and praxis. In J. Dindyal, L.P. Cheng, & S.F. Ng (Eds.) *Mathematics education: Expanding horizons, (Proceedings of the 35th annual conference of the Mathematics Education Research Group of Australasia, eBook* (pp. 618–624). Singapore: MERGA.

Dindyal, J., Quek, K.S., Leong, Y.H., Toh, T.L., Tay, E.G., & Lou, S.T. (2010). Problems for a problem solving curriculum. In L. Sparrow, B. Kissane, & C. Hurst (Eds.) *Shaping the future of mathematics education* (pp. 749–752). Fremantle: Australia.

Leong, Y.H., Toh, T.L., Quek, K.S., Dindyal, J., & Tay, E.G. (2010). Enacting a problem solving curriculum. In L. Sparrow, B. Kissane, & C. Hurst (Eds.) *Shaping the future of mathematics education* (pp. 749–752). Fremantle: Australia.

Quek, K.S., Toh, T.L., Dindyal, J., Leong, Y.H., Tay, E.G., & Lou, S.T. (2010). Resources for Teaching problem solving: a problem to discuss. In L. Sparrow, B. Kissane, & C. Hurst (Eds.) *Shaping the future of mathematics education* (pp. 749–752). Fremantle: Australia.

Leong, Y.H., Toh, T.L., Quek, K.S., Dindyal, J., & Tay, E.G. (2009). Teacher preparation for a problem solving curriculum. In R. Hunter, B. Bicknell, & T. Burgress (Eds.), *MERGA 32 Conference proceedings* (pp. 691–694). Wellington, New Zealand: Massey University.

Dindyal, J., Toh, T.L., Quek, K.S., Leong, Y.H., & Tay, E.G. (2009). Reconceptualizing problem solving in the school curriculum. In R. Hunter, B. Bicknell, & T. Burgress (Eds.), *MERGA 32 Conference proceedings* (pp. 691–694). Wellington, New Zealand: Massey University.

Toh, T.L., Quek, K.S., Leong, Y.H., Dindyal, J., & Tay, E.G. (2009). Assessment in a problem solving curriculum In R. Hunter, B. Bicknell, & T. Burgress (Eds.), *MERGA 32 Conference proceedings* (pp. 691–694). Wellington, New Zealand: Massey University.